T0257964

Diverse Applications of Wavelet Theory

Diverse Applications of Wavelet Theory

Edited by **Victor Nason**

CLANRYE INTERNATIONAL

New Jersey

Published by Clanrye International,
55 Van Reypen Street,
Jersey City, NJ 07306, USA
www.clanryeinternational.com

Diverse Applications of Wavelet Theory
Edited by Victor Nason

© 2015 Clanrye International

International Standard Book Number: 978-1-63240-151-9 (Hardback)

Contents

Preface

Every book is a source of knowledge and this one is no exception. The idea that led to the conceptualization of this book was the fact that the world is advancing rapidly; which makes it crucial to document the progress in every field. I am aware that a lot of data is already available, yet, there is a lot more to learn. Hence, I accepted the responsibility of editing this book and contributing my knowledge to the community.

This book elaborates on Wavelet Theory and its applications in various fields. The application of wavelet transformation to examine the behavior of complex systems from several fields has begun to be widely identified and applied successfully during the past few decades. The book deals with three major aspects of Wavelet Transformation i.e. Electrical Systems, Fault Diagnosis and Monitoring, and Signal Processing. One of the important characteristics of this book is that the wavelet concepts that are applied in engineering, physics and technology have been discussed from a point of view that is recognizable to researchers from different branches of science and engineering. The text of the book is worth utility to a large number of readers.

While editing this book, I had multiple visions for it. Then I finally narrowed down to make every chapter a sole standing text explaining a particular topic, so that they can be used independently. However, the umbrella subject sinews them into a common theme. This makes the book a unique platform of knowledge.

I would like to give the major credit of this book to the experts from every corner of the world, who took the time to share their expertise with us. Also, I owe the completion of this book to the never-ending support of my family, who supported me throughout the project.

Editor

Part 1

Image Processing

The Wavelet Transform
for Image Processing Applications

Bouden Toufik[1] and Nibouche Mokhtar[2]

[1]*Automatic Department, Laboratory of Non Destructive Testing, Jijel University*
[2]*Bristol Robotic Laboratory, Department of Electrical and Computer Engineering,*
University of the West of England
[1]*Algeria*
[2]*UK*

1. Introduction

In recent years, the wavelet transform emerged in the field of image/signal processing as an alternative to the well-known Fourier Transform (FT) and its related transforms, namely, the Discrete Cosine Transform (DCT) and the Discrete Sine Transform (DST). In the Fourier theory, a signal (an image is considered as a finite 2-D signal) is expressed as a sum, theoretically infinite, of sines and cosines, making the FT suitable for infinite and periodic signal analysis. For several years, the FT dominated the field of signal processing, however, if it succeeded well in providing the frequency information contained in the analysed signal; it failed to give any information about the occurrence time. This shortcoming, but not the only one, motivated the scientists to scrutinise the transform horizon for a "messiah" transform. The first step in this long research journey was to cut the signal of interest in several parts and then to analyse each part separately. The idea at a first glance seemed to be very promising since it allowed the extraction of time information and the localisation of different frequency components. This approach is known as the Short-Time Fourier Transform (STFT). The fundamental question, which arises here, is how to cut the signal? The best solution to this dilemma was of course to find a fully scalable modulated window in which no signal cutting is needed anymore. This goal was achieved successfully by the use of the wavelet transform.

Formally, the wavelet transform is defined by many authors as a mathematical technique in which a particular signal is analysed (or synthesised) in the time domain by using different versions of a dilated (or contracted) and translated (or shifted) basis function called the wavelet prototype or the mother wavelet. However, in reality, the wavelet transform found its essence and emerged from different disciplines and was not, as stated by Mallat, totally new to mathematicians working in harmonic analysis, or to computer vision researchers studying multiscale image processing (Mallat, 1989).

At the beginning of the 20th century, Haar, a German mathematician introduced the first wavelet transform named after him (almost a century after the introduction of the FT, by the French J. Fourier). The Haar wavelet basis function has compact support and integer coefficients. Later, the Haar basis was used in physics to study Brownian motion (Graps,

1995). Since then, different works have been carried out either in the development of the theory related to the wavelet, or towards its application in different fields. In the field of signal processing, the great achievements reached in different studies by Mallat, Meyer and Daubechies have allowed the emergence of a wide range of wavelet-based applications. In fact, inspired by the work developed by Mallat on the relationships between the Quadrature Mirror Filters (QMF), pyramid algorithms and orthonormal wavelet bases (Mallat, 1989), Meyer constructed the first non-trivial wavelets (Meyer, 1989). However, the most important work was carried out by Ingrid Daubechies. Based on Mallat's work, Daubechies succeeded to construct a set of wavelet orthonormal basis functions, which have become the cornerstone of many applications (Daubechies, 1988). Few years later, the same author, in collaboration with others (Cody, 1994), presented a set of wavelet biorthogonal basis function, which later found their use in different applications, especially in image coding. Recently, JPEG2000, a biorthogonal wavelet-based compression has been adopted as the new compression standard (Ebrahimi et al., 2002).

2. Continuous Wavelet Transform

Different ways to introduce the wavelet transform can be envisaged (Starck et al., 1998). However, the traditional method to achieve this goal remains the use of the Fourier theory (more precisely, STFT). The Fourier theory uses sine and cosine as basis functions to analyse a particular signal. Due to the infinite expansion of the basis functions, the FT is more appropriate for signals of the same nature, which generally are assumed to be periodic. Hence, the Fourier theory is purely a frequency domain approach, which means that a particular signal f(t) can be represented by the frequency spectrum F(w), as follows:

$$F(\omega) = \int_{-\infty}^{+\infty} f(t)e^{-j\omega t}dt \tag{1}$$

The original signal can be recovered, under certain conditions, by the inverse Fourier Transform as follows:

$$f(t) = \frac{1}{2\pi}\int_{-\infty}^{+\infty} F(\omega)e^{j\omega t}d\omega \tag{2}$$

Obviously, discrete-time versions of both direct and inverse forms of the Fourier transform are possible.

Due to the non-locality and the time-independence of the basis functions in the Fourier analysis, as represented by the exponential factor of equation (1), the FT can only suit signals with "time-independent" statistical properties. In other words, the FT can only provide global information of a signal and fails in dealing with local patterns like discontinuities or sharp spikes (Graps, 1995). However, in many applications, the signal of concern is both time and frequency dependent, and as such, the Fourier theory is "incapable" of providing a global and complete analysis. The shortcomings of the Fourier transform, in addition to its failure to deal with non-periodic signals led to the adoption by the scientific community of a windowed version of this transform known as the STFT. The STFT transform of a signal f(t) is defined around a time θ through the usage of a sliding window w (centred at time θ) and a frequency ω as (Wickerhauser, 1994; Graps, 1995; Burrus et al., 1998; David, 2002 & Oppenheim & Schafer, 2010):

$$\text{STFT}(\theta, w) = \int_{-\infty}^{+\infty} f(t)w(t-\theta)e^{-jwt}dt \tag{3}$$

As it is apparent from equation (3), even if the integral limits are infinite, the analysis is always limited to a portion of the signal, bounded by the limits [-θ, θ] of the sliding window. The time-frequency plane of a fixed window STFT transform is illustrated in Figure 1.

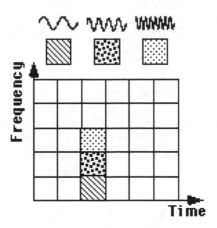

Fig. 1. Fourier time-frequency plane (Graps, 1995)

Although, this approach (using STFT transform) succeeds well in giving both time and frequency information about a portion of the signal, however, as its predecessor, it has a major drawback. The fact is that the choice of the window size is crucial. As stated by Starck and al (Starck et al., 1998): " The smaller the window size, the better the time-resolution. However, the smaller the window size also, the more the number of discrete frequencies which can be represented in the frequency domain will be reduced, and therefore the more weakened will be the discrimination potential among frequencies". This problem is closely linked to the Heisenberg's uncertainty principle, which states that a signal (e.g. a very short portion of the signal) cannot be represented as a point in the time-frequency domain.

This shortcoming brings us to rise the fundamental question: how to size then the sliding window? Not surprisingly, the answer to this question leads us by means of certain transformations to the wavelet transform. In fact, by considering the convolution of the sliding window with the time-dependant exponential e^{-jwt} within the integral of equation (3):

$$K_{\theta,\omega}(t) = w(t-\theta)e^{-jwt} \tag{4}$$

And replacing the frequency ω by a scaling factor a, and the window bound θ by a shifting factor b, leads us to the first step leading to the Continuous Wavelet Transform (CWT), as represented in equation (5):

$$K_{a,b}(t) = \frac{1}{\sqrt{a}}\psi^*(\frac{t-b}{a}) \qquad a \in R^+, b \in R \tag{5}$$

The combination of equation (5) with equation (3), leads to the CWT as defined by Morlet and Grossman (Grossman & Morlet, 1984).

$$W(a,b) = \frac{1}{\sqrt{a}} \int_{-\infty}^{+\infty} f(t) \psi^* \left(\frac{t-b}{a} \right) dt \qquad (6)$$

Where f(t) belongs to the square integrable functions space, $L^2(R)$. In the same way, the inverse CWT can be defined as (Grossman & Morlet, 1984):

$$f(t) = \frac{1}{C_\psi} \int_0^{+\infty} \int_{-\infty}^{+\infty} \frac{1}{\sqrt{a}} W(a,b) \psi \left(\frac{t-b}{a} \right) \frac{da\,db}{a^2} \qquad (7)$$

The C_ψ factor is needed for reconstruction purposes. In fact, the reconstruction is only possible if this factor is defined. This requirement is known as the admissibility condition. In a more general way, $\psi(t)$ is replaced by $\chi(t)$, allowing a variety of choices, which can enhance certain features for some particular applications (Starck et al., 1998; Stromme, 1999 & Hankerson et al., 2005). However, the CWT in the form defined by equation (6) is highly redundant, which makes its direct implementation of minor interest. The time-frequency plane of a wavelet transformation is illustrated in Figure 2. The differences with the STFT transform are visually clear.

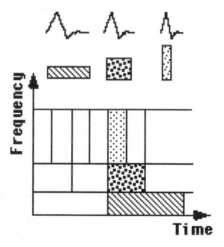

Fig. 2. Wavelet time-frequency plane ((Graps, 1995) with minor modifications)

At this stage and after this brief introduction, it is natural to ask the question: therefore what are wavelet Transforms?

Although wavelet transforms are defined as a mathematical tool or technique, there is no consensus within the scientific community on a particular definition. This "embarrassment" has been stated by Sweldens as (Sweldens, 1996): "Giving that the wavelet field keeps growing, the definition of a wavelet continuously changes. Therefore it is impossible to rigorously define a wavelet". According to the same author, to call a particular function a wavelet system, it has to fulfil the three following properties:

- Wavelets are building blocks for general functions: They are used to represent signals and more generally functions. In other words, a function is represented in the wavelet space by mean of infinite series of wavelets.
- Wavelets have space – frequency localisation: Which means that most of the energy of a wavelet is confined in a finite interval and that the transform contains only frequencies from a certain frequency band.
- Wavelets support fast and efficient transform algorithms: This requirement is needed when implementing the transform. Often wavelet transforms need O(n) operations, which means that the number of multiplications and additions follows linearly the length of the signal. This is a direct implication of the compactness property of the transform. However, more general wavelet transforms require O(nlog(n)) operations (e.g. undecimated wavelet).

To refine the wavelet definition, the three following characteristics have been added by Sweldens and Daubechies (Sweldens, 1996 & Daubechies, 1992, 1993) as reported in (Burrus et al., 1998):

- Oneness of the generating function: Refers to the ability of generating a wavelet system from a single scaling function or wavelet function just by scaling and translating.
- Multiresolution ability: This concept, which has first been introduced by Mallat, states the ability of the transform to represent a signal or function at different level, by different weighted sums, derived from the original one.
- Ability of generating lower level coefficients from the higher level coefficients. This can be achieved through the use of tree-like structured chain of filters called Filter Banks.

3. Multiresolution

The multiresolution concept has been introduced first by Mallat (Mallat, 1989). It defines clearly the relationships between the QMF, pyramid algorithms and orthonormal wavelet bases through basically, the definition of a set of nested subspaces and a so-called scaling function. The strength of multiresolution lies in its ability to decompose a signal in finer and finer details. Most importantly, it allows the description of a signal in terms of time-frequency or time-scale analysis.

3.1 Nested subspaces

The basic requirement for multiresolution analysis is the existence of a set of approximation subspaces of $L^2(R)$ (square integrable function space) with different resolutions, as represented schematically for the three intermediate subspaces in Figure 3 and stated by equation (8):

$$V_{-\infty}....\subset V_{-1} \subset V_0 \subset V_1 \subset ....V_\infty = L^2(R) \tag{8}$$

In such a way that, if $f(t) \in V_j$ then $f(2t) \in V_{j+1}$. Which means that the subspace containing high resolution will automatically contains those of lower resolution. In a more general case, if $f(t) \in V_0$, then $f(2^k t) \in V_k$. This implication is known as the scale invariance property.

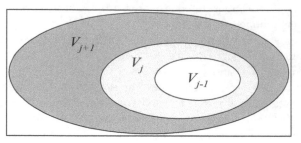

Fig. 3. Nested subspaces

3.2 Scaling function

The existence of a so-called scaling function $\phi(t)$ is primordial in order to benefit from the multiresolution concept. In this context, let us define the scaling function first and then define the wavelet function through it (Burrus et al., 1998). Let the scaling function be defined by the following equation:

$$\varphi_k(t) = \varphi(t-k) \qquad k \in Z \qquad \varphi \in L^2(R) \tag{9}$$

Which forms with its translates an orthonormal (The orthogonality is not necessary, since a non orthogonal basis (with the shift property) can always be orthogonalised (Sweldens, 1995)) basis of the space V_0:

$$V_0 = \operatorname*{span}_k\{\varphi_k(t)\} \tag{10}$$

This means that any function belonging to this space ($f(t) \in V_0$) can be expressed as a linear combination of a set of so-called expansion coefficients, with the scaling function and its consecutive translates (since $\varphi_k(t)$ are the basis functions):

$$f(t) = \sum_k c_k \varphi_k(t) = \sum_k c(k)\varphi(t-k) \tag{11}$$

Where the expansion coefficients c_k (or $c(k)$) are calculated using the inner product:

$$c_k = \langle f(t), \varphi_k(t) \rangle \tag{12}$$

By simply scaling and translating, a two-dimensional scaling function is generated from the original scaling function defined in equation (9):

$$\varphi_{j,k}(t) = \frac{1}{\sqrt{a}}\varphi(\frac{t-b}{a}) \tag{13}$$

Where a and b are, the scaling and the shifting factors as defined in equation (5), respectively. To ease the implementation of a wavelet system, the translation and the scaling factor have been adopted to be a factor of two. In fact (Graps, 1995):

$$a = 2^{-j} \quad , \quad b = 2^{-j}.k \tag{14}$$

These values are adopted for the remaining of the chapter. Thus equation (13) can be rewritten as:

$$\varphi_{j,k}(t) = 2^{j/2}\,\varphi(2^{j}t - k) \tag{15}$$

Identically, the two-dimensional scaling function forms with its translates an orthonormal space over k:

$$V_j = \operatorname*{span}_{k}\{\varphi_{j,k}(t)\} \quad k \in Z \quad \text{and} \quad j \in Z \tag{16}$$

And as such any function f(t) of this space can be expressed as:

$$f(t) = \sum_{k} c(k)\varphi(2^{j}t + k) \tag{17}$$

As a consequence, if $\varphi(t) \in V_0$, then since $V_0 \subset V_1$, $\varphi(t)$ can be expressed as a linear combination of the scaling function $\varphi(2t)$ spanning the space V_1:

$$\varphi(t) = \sum_{k} h(k)\sqrt{2}\,\varphi(2t - k) \tag{18}$$

Where the coefficients h(k) are the scaling function coefficients. The value $\sqrt{2}$ ensures that the norm of the scaling function is always equal to the unity. This equation is fundamental to the multiresolution theory and is called the multiresolution analysis equation.

4. Wavelet function

What has been done so far to define the scaling function, its translates and the corresponding spanned spaces, can also be applied in the same way to the so-called wavelet function. Let us suppose for this purpose that the subspace $V_0 \subset V_1$ has an orthogonal complement W_0, such as V_1 can be represented as a combination of V_0 and W_0 as follows:

$$V_1 = V_0 \oplus W_0 \tag{19}$$

Where the complementary space W_0 is spanned also by an orthonormal basis:

$$\psi_k(t) = \psi(t - k) \quad k \in Z \qquad \psi \in L^2(R) \tag{20}$$

The function $\psi(t)$ is known as the mother wavelet, the wavelet prototype or the wavelet function. The same properties, which apply to the scaling function, are also applicable to the wavelet function. In other words, a function $f(t) \in W_0$ can be expressed as:

$$f(t) = \sum_k d_k \, \psi_k(t) = \sum_k d(k)\psi(t-k) \qquad (21)$$

Where, the expansion coefficients d_k (or $d(k)$) are calculated using the inner product:

$$d_k = <f(t), \psi_k(t)> \qquad (22)$$

Likewise, since $W_0 \subset V_1$, $\psi(t)$ can also be expressed in terms of the scaling function $\varphi(2t)$ of the higher space V_1:

$$\psi(t) = \sum_k g(k)\sqrt{2}\, \varphi(2t-k) \qquad (23)$$

Where g(k) are the wavelet coefficients. This leads to a dyadic decomposition as represented by the grid of Figure 5. The equation (19) can be generalised to an arbitrary number of subspaces, such as, V_2 is represented in terms of V_1 and W_1, V_3 in terms of V_2 and W_2, and so on. The whole decomposition process is illustrated in Figure 4.

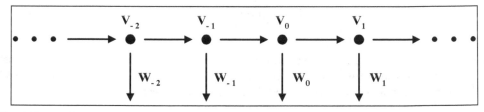

Fig. 4. Space decomposition

More generally, a subspace V_j is spanned by W_{j-1} and V_{j-1}. Thus, the $L^2(R)$ space can be decomposed as follows:

$$L^2(R) = V_j \oplus W_j \oplus W_{j+1} \oplus W_{j+2} \oplus \ldots W_0 \oplus W_1 \ldots \qquad (24)$$

The index j represents the depth or the level of decomposition, which is arbitrary in this case. As for the scaling function, a two-dimensional scaled and translated wavelet function is defined as:

$$\psi_{j,k}(t) = 2^{j/2}\, \psi(2^j t - k) \qquad (25)$$

In such way that:

$$W_j = \operatorname*{span}_k \{\psi_{ji,k}(t)\} \qquad (26)$$

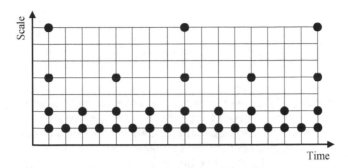

Fig. 5. Dyadic wavelet transform space representation

5. Series expansions and Discrete Wavelet Transforms

According to equation (24), a function $f(t)$ belonging to the $L^2(R)$ space can be expanded in series in terms of the scaling function spanning the space V_j and the wavelet functions spanning the spaces W_j, W_{j+1}, W_{j+2},....., W_0, W_1.... as follows:

$$f(t) = \sum_k c_j(k)\varphi_{j,k}(t) + \sum_{n=j}^{+\infty}\sum_{k=-\infty}^{+\infty} d_n(k)\psi_{n,k}(t) \qquad (27)$$

Where $\varphi_{j,k}(t)$ is defined by equation (15) and $\psi_{n,k}(t)$ is defined by equation (25). In this case, the index j, which is arbitrary, represents the coarsest scale, while the remaining are the high resolution details. Equation (27) represents the wavelet expansion series of the function $f(t)$, which plays a major role when deriving a more practical form of the wavelet transform.

The coefficients in the wavelet expansion series $c_j(k)$ and $d_n(k)$ (or $c(j,k)$ and $d(n,k)$) are the so-called discerete wavelet transform of the function $f(t)$. Since the basis functions are orthonormal, they can be calculated using equations (12 and 22), respectively. We will see later in this chapter that the orthonormality condition can be relaxed allowing the implementation of another important basis, namely, the biorthogonal basis.

6. Filter banks and wavelet implementations

In general, wavelet transform-based applications involve discrete coefficients instead of scaling and/or wavelet functions. For practical and computational reasons, discrete time filter banks are required. Such structures decompose a signal into a coarse representation along with added details. To achieve this representation, the relationship between the expansion coefficients at lower and higher scale levels need to be defined. This can be easily done by using a scaled and shifted version of equation (18) along with simple transformations as reported in (Burrus et al., 1998). This relation is defined by:

$$c_j(k) = \sum_n h(n-2k)c_{j+1}(n) \qquad (28)$$

And

$$d_j(k) = \sum_n g(n - 2k)c_{j+1}(n) \tag{29}$$

Where $n \in Z$ and $k \in Z$. The computation of such equations is achieved through the use of the well-established digital filtering theory. In particular, for finite length signals (which is the case for digital images), the use of a Finite Impulse Response filter (FIR) is the most appropriate choice. However, since equations (28 and 29) compute one output for each two consecutive inputs, a modification needs to be made. The basic operation required here, is derived from the multirate signal processing theory (Fliege, 1994; Hankerson et al., 2005; Cunha et al., 2006; Lu & Do, 2007; Nguyen & Oraintara, 2008 & Brislawn, 2010). It simply consists of using a down-sampler or decimator by a factor of two. In practice, it consists of applying a pair of FIR filters; each followed by a decimator as illustrated by Figure 6:

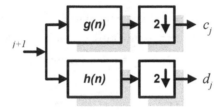

Fig. 6. Analysis Filter Bank

The filter bank is defined as a combination of a low pass filter and high pass filter, both followed by a factor of two decimation (Strang & Nguyen, 1996). Thus, the decomposition is reduced to two basic operations from the digital signal processing theory: a filtering and a down sampling.

The structure in Figure 6 is generally used to implement Mallat's algorithm. To allow further level of decomposition, identical stages are cascaded leading to a multiresolution analysis. This analysis scheme is known as the Subband Coding structure (Burrus et al., 1998) and is illustrated in the following figure.

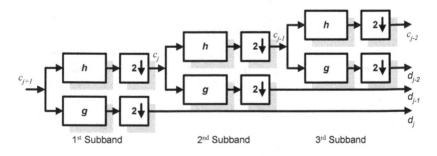

Fig. 7. Three-Stage analysis Subband Coding

At each stage, the spectrum frequency of the analysed signal is halved by a factor of two. This leads to a logarithmic set of bandwidths as illustrated by Figure 8.

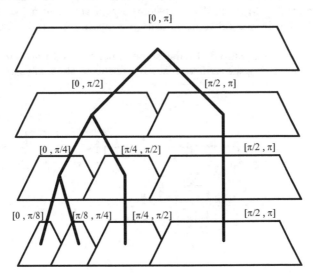

Fig. 8. Frequency Spectrum of a three-stage Subband Structure

To recover the original signal from the previously analysed one, a reversed version of the analysis filter bank of Figure 6 is required. This can be achieved by using two basic operations: a filtering and an up sampling or interpolating process. In multirate digital signal processing, appending a zero sample between two consecutive samples performs the up sampling. Thus, for each input sample, we get two output samples. A three-stage synthesis subband coding is illustrated in Figure 9.

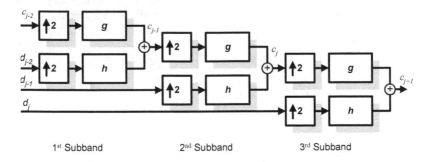

Fig. 9. Three-Stage synthesis Subband Coding

7. Algorithms for Wavelet Transform computation

This section is concerned with a review of variety of algorithms dedicated to implement wavelet transforms. We focus on both 1-Dimensional and 2-Dimensional systems.

7.1 Burt's Pyramid

Dedicated initially to lossless image coding, the pyramid algorithm was first introduced by Burt (Burt & Adelson, 1983). Basically, it decomposes a signal in a low-resolution signal along with some higher resolution signals through a repetition of reduction and expansion processes. At each level, the reduced and expanded signal is compared with the original signal and the difference is stored. In the same time, the reduced signal is repeatedly decomposed by further using the reducer block in the chain. The analysis/synthesis process is shown in Figure 10.

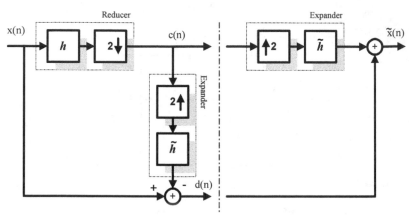

Fig. 10. Pyramidal analysis and synthesis

The reduction block performs the two basic operations of a low pass filtering and decimating by a factor of 2. The expansion block up samples the signal first, then filters it through the use of a synthesis low pass filter. To reconstruct the original signal, the difference signal at each level is added to a previously expanded signal. Repeatedly, the resulting signal is expanded and added to the corresponding difference signal.

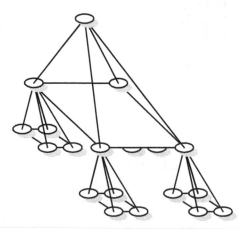

Fig. 11. 2-D Pyramidal Structure

The decomposition and the reconstruction processes for a 2-D signal, as in image processing, is achieved through the use of a 2-D filtering process. In this case, only 1/4 of the original signal is obtained at the output of the reducer (the decimation is performed twice). This scheme can be represented by the pyramidal structure of Figure 11.

This type of decomposition makes this algorithm suitable for a progressive image transmission scheme.

7.2 Mallat's Pyramidal algorithm

Mallat's pyramid is a direct consequence of the multiresolution concept developed by the same author and presented in section 6. Up to date, it is the most widely used approach - both in software and hardware - for implementing the wavelet transform (Masud, 1999). Since the one-dimensional decomposition and reconstruction schemes have been already introduced in section 6, we will focus in this section on two-dimensional schemes, which are more suitable for image analysis and synthesis. The two-dimensional decomposition approach is based on the property of separation of the functions into arbitrary x and y directions. The first step is identical to the one-dimensional approach, however, instead of keeping the low-level resolution and processing the high level resolution, both are processed using two identical filter bank after a transposition of the incoming data. Thus, the image is scanned in both horizontal and vertical directions. This result in an average image (or subimage) and three detail images generated by the following 2-D scaling function $\varphi(x,y) = \varphi(x)\varphi(y)$ and the vertical, the horizontal and the diagonal wavelet functions: $\psi_1(x,y) = \varphi(x)\psi(y)$, $\psi_2(x,y) = \psi(x)\varphi(y)$ and $\psi_3(x,y) = \psi(x)\psi(y)$, respectively. To recover the original image, the inverse process is applied. Figure 12 illustrates the analysis and synthesis stages built using three filter banks each.

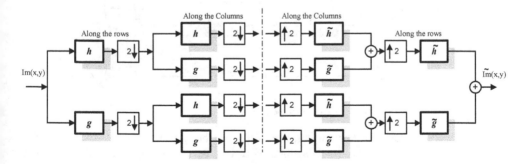

Fig. 12. Two-dimensional Mallat's analysis and synthesis tree

In this case, the frequency band is halved at each stage by a factor of four as represented by Figure 13.

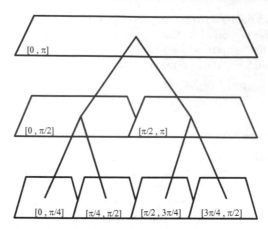

Fig. 13. Frequency Bands of Mallat's 2-D Analysis Algorithm

7.3 Feauveau's non-dyadic structure

Based on Adelson's work (Adelson et al., 1987), this approach has been introduced by Feauveau (Feauveau, 1990). This decomposition is also known as Quincux. It differs from Mallat's two-dimensional approach by the fact that only the decimated output from the low pass filter is transposed and then processed through a "similar" filter bank. The result is a low resolution average image along with two different detail images from two different resolution levels. The fact is that the decomposition is not dyadic and the initial resolution of a factor of 2 is replaced by a $\sqrt{2}$ factor leading to an asymmetrical support. Figure 14 shows an analysis and synthesis stage of a Quincux structure.

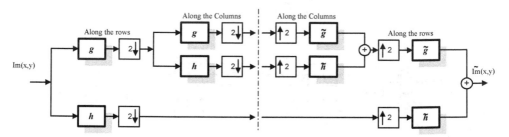

Fig. 14. Feauveau's analysis and synthesis tree

Due to the removal of the filter bank at the output of the high pass filter - as reported in (Starck et al., 1998) only a wavelet image is involved at each stage. Recently, this approach has been used in an image compression scheme and found to give often better overall performances than other approaches (Stromme, 1999; Ebrahimi et al, 2002; Smith, 2003; Hankerson et al., 2005; Xiong & Ramchandran, 2005; Nai-Xiang et al., 2006; Raviraj & Sanavullah, 2007 & Oppenheim & Schafer, 2010). The frequency bands of a Quincux analysis is shown in Figure 15.

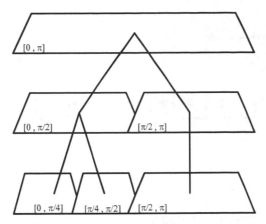

Fig. 15. Frequency bands of Feauveau's Quincux decomposition

7.4 Swelden's lifting scheme

Unlike the three previous methodologies, the lifting scheme follows another philosophy. The fact is that the Fourier theory is not involved anymore and the construction of any wavelet system lies only in the spatial domain. If the explanation of the theory relies on the works of Sweldens (Sweldens, 1995, 1996 & Valens, 2004) the lifting approach has links with many other schemes (Burrus et al., 1998; Do & Vetterli, 2003, 2005; Cunha et al., 2006; Lu & Do, 2007; Nguyen & Oraintara, 2008 & Brislawn, 2010). The lifting-based wavelet transform can be seen as a succession of three operations: split, predict and update. In the first operation, data is the split into even and odds parts (known also as the lazy wavelet transform). Then, differences or details are calculated through the usage of a predictor. Finally, to compute the average, the even part is updated using the details previously calculated. Figure 16 shows an analysis and synthesis lifting-based wavelet transform.

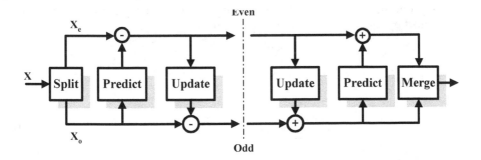

Fig. 16. Lifting-based Wavelet Transform

The reconstruction operation does exactly the same, but using the reverse process. The data is first predicted, then updated and finally merged. Figure 17 illustrates split and merge operations using the polyphase property (Fliege, 1994).

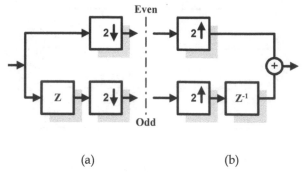

(a)   (b)

Fig. 17. Lazy Wavelet Transform: (a) Split, (b) Merge

8. The Wavelet Transform revisited

In many practical problems, both the orthonormal basis (Daubechies, 1988, 1992, 1993) and the biorthogonal basis (Cody, 1994) can be used. The two bases (or families) present similarities and differences. Another scheme, called wavelet packet, which involves either orthonormal basis or biorthogonal basis is also possible (Wickerhauser, 1994). The following briefly describes the main features of orthonormal and biorthogonal bases together with extension to the wavelet packet scheme. It is worth mentioning that other schemes like undecimated wavelet, adaptive wavelets and multiwavelets exist and are beyond the scope of this brief overview.

8.1 Orthonormal basis

The orthonormal basis emerged from the work initiated by Mallat and Daubechies (Mallat, 1989 & Daubechies, 1988, 1993). The orthonormality property is somewhat seen as the discrete version of the orthogonality property (Masud, 1999). However, the basis functions are further normalised. These concepts have been mentioned when the multiresolution feature and the scaling function have been introduced. The admissibility and the orthogonality conditions ensure the existence and the orthogonality feature of the scaling function, defined by equation (18). This is achieved if:

$$\sum_n h(n) = \sqrt{2} \tag{30}$$

And

$$\sum_n h(n)h(n + 2k) = \delta(k) \tag{31}$$

Furthermore, using the two equations above alongside with equation (23), which defines the wavelet function, the orthogonality of the scaling function and the wavelet function at the same scale can be derived. This can be achieved only if the following equality is verified:

$$g(n) = (-1)^n h(1 - n) \tag{32}$$

The orthogonality between the wavelet coefficients and the scaling coefficients is then only a simple implication:

$$\sum_n h(n)g(n) = 0 \tag{33}$$

The scaling coefficients, which satisfy equation (33), are called Quadrature Mirror Filters (QMF).

To achieve perfect reconstruction, the analysed signal has to be identical to the synthesised one. In other words, $c_j(n) = \tilde{c}_j(n)$, where $c_j(n)$ and $\tilde{c}_j(n)$ are the input and the output of a two-band filter (or filter bank) as shown in Figure 18, respectively.

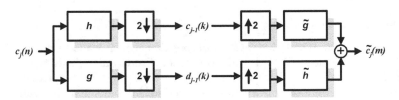

Fig. 18. Two-band analysis and synthesis filter bank

8.2 Biorthogonal basis

Biorthogonal wavelet basis can be seen as a generalisation of the orthogonal wavelet basis where some imposed restrictions on the latter have been relaxed. Unlike the case of orthogonal basis, the scaling and the wavelet functions need be neither of the same length, nor even numbered. Hence, the quadrature mirror property is not applicable and is replaced with a dual property. For the perfect reconstruction equation to hold, the scaling and the wavelet coefficients have to fulfil the following equations:

$$\tilde{g}(n) = (-1)^n h(1-n) \tag{34}$$

$$g(n) = (-1)^n \tilde{h}(1-n) \tag{35}$$

It is clear that when the analysis and the synthesis filters are similar, the system becomes orthogonal. The "orthogonality" condition in this case is defined by:

$$\sum_n \tilde{h}(n)h(n+2k) = \delta(k) \tag{36}$$

Previously, in orthogonal basis, only the analysis scaling coefficients (or wavelet coefficients) along with their shifted versions were used. In biorthogonal case, the analysing scaling coefficients are kept unchanged, while their shifted versions are replaced by the shifted versions of the synthesis dual filter. In other words, the analysis filter is orthogonal to its synthesis dual filter. The biorthogonal denomination comes from this feature.

At the expense of the energy partitioning property stated by Perseval's equality, which is a direct consequence of the lack of orthogonality, a greater flexibility can be achieved by using the basis and dual basis (Burrus et al., 1998). One of the most "important" features in the biorthogonal basis is the linear phase property, which leads to the filter coefficients (when implementing a wavelet system) being symmetric. In addition, the difference of length between dual filters must be even, leading either to odd or even length of the low pass and the high pass filters. In general, biorthogonal wavelet systems present the following features (Daubechies, 1992):

- The coefficients of the filters are either real numbers or integers;
- The filters in this family present either even or odd orders;
- The low pass and the high pass filters used in the filter bank have not the same length;
- The low pass filter is always symmetric;
- The high pass filter is either symmetric or antisymmetric.

8.3 Wavelet packets

In contrast to the "traditional" Mallat's decomposition, which leads to narrow frequency bandwidths (low frequencies) and wide frequency bandwidths (high frequencies), the wavelet packet approach emerged first as a way of adjusting high frequency resolutions. Hence, the Mallat's decomposition scheme is applied to both parts of a filter bank leading to the split of frequencies in progressive finer resolutions. The generic structure of wavelet packet decomposition is shown in Figure 19 and the frequency bandwidths illustrated by Figure 20. In this scheme, the number of filters increases by a factor of $(2^i - 2^j)$ at each successive subband, where i and j represent two consecutive resolutions and $i - j = 1$.

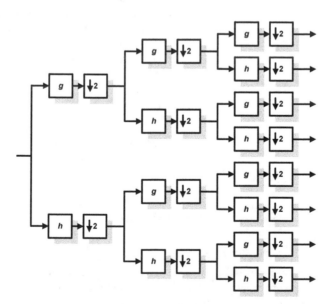

Fig. 19. Three-stage Wavelet Packet Decomposition

In comparison to classical wavelet approach, the wavelet packet scheme presents the following features (Daubechies, 1992):

- Possibility of using different wavelet from a level to another. This strategy has been used in (Masud, 1999) to implement a two-level orthonormal wavelet packet and a three-level biorthogonal wavelet packet.
- Possibility of choosing a particular wavelet packet decomposition from the general generic structure of Figure 19. Thus, one can choose either to preserve the orthonormality feature of the decomposition (Wickerhauser, 1994), or highlight the peculiarities of the signal (Masud, 1999). A binary search for the best decomposition tree is also possible (Burrus et al., 1998).

However, there is a cost to be paid. In this case, the computational complexity of a wavelet packet structure is $O(n\log(n))$ in contrast to the $O(n)$ of the classical wavelet transform.

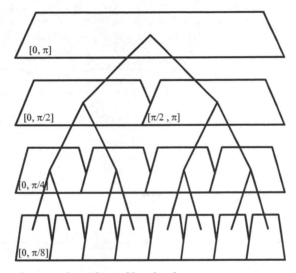

Fig. 20. Two-band analysis and synthesis filter bank

9. Wavelet-based applications

Recently, The wavelet transform is being increasingly used, not only in the field of image and signal processing applications but also in many other different areas, ranging from mathematics, physics, astronomy to statistics and economics. In image processing based applications, image compression, image denoising and image watermarking are at the cutting edge, and as such, a brief description of these wavelet-based applications is given in the following subsections (Strang & Nguyen, 1996; Burrus et al., 1998; Stromme, 1999; Ebrahimi et al, 2002; Nibouche et al., 2000, 2001a, 2001b, 2001c, 2001d, 2002, 2003; Smith, 2003; Do & Vetterli, 2003, 2005; Hankerson et al., 2005; Nai-Xiang Yap-Peng, 2005; Xiong & Ramchandran, 2005; Chappelier & Guillemot, 2006; Cunha et al., 2006; Nai-Xiang et al., 2006; Raviraj & Sanavullah, 2007; Hernandez-Guzmane et al., 2008; Firoiu et al., 2009; Mallat, 2009; Brislawn, 2010; Oppenheim & Schafer, 2010; Ruikar & Doye, 2010 & Chen & Qian, 2011).

9.1 Image compression

Even though the wavelet transforms have been widely used in image coding since the late 80s, they only gained their notoriety in the field by the adoption of the first wavelet-based compression standard scheme, known as the FBI fingerprint compression standard Bradley, et al., 1993). Recently, what did Sweldens state in (Sweldens, 1996) as a need of standardising a wavelet-based compression scheme under the header "problems not sufficiently explored with wavelets", has seen the day, by the adoption of the JPEG2000 new compression standard (Ebrahimi et al., 2002). The block diagram of the JPEG2000 standard does not really differ from the JPEG standard one. The discrete wavelet transform, which replaces the DCT, is applied first to the source image. The transformed coefficients are then quantised. Finally, the output coefficients from the quantiser are encoded (using either Huffman coding or arithmetic coding techniques) to generate the compressed image (Smith, 2003; Do & Vetterli, 2005; Hankerson et al., 2005; Xiong & Ramchandran, 2005; Chappelier & Guillemot, 2006; Nai-Xiang et al., 2006; Raviraj & Sanavullah, 2007; Mallat, 2009; Oppenheim & Schafer, 2010). To recover the original image the inverse process is applied. Figure 21 shows the basic JPEG2000 Encoding Scheme (Ebrahimi et al., 2002).

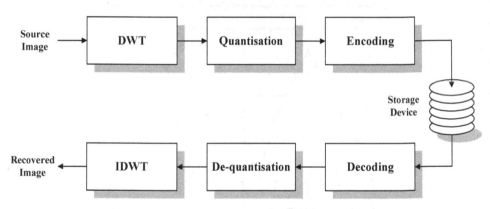

Fig. 21. Wavelet-based encoding scheme

9.2 Image denoising

Image manipulation, includes a wide range of operations like digitising, copying, transmitting, displaying ... etc. Unfortunately, such manipulations generally degrade the image quality by spanning many types of noise. Hence, to recover the original structure of the image, the undesired added noise needs to be localised and then removed. In image processing, noise removal is achieved through the usage of filtering-based denoising techniques (Nai-Xiang & Yap-Peng, 2005; Chappelier & Guillemot, 2006; Firoiu et al., 2009; Mallat, 2009; Nafornita et al., 2009; Ruikar & Doye, 2010; Oppenheim & Schafer, 2010 & Chen & Qian, 2011). Traditionally, image denoising or image enhancement is performed using either linear filtering or non-linear filtering. Linear filtering is achieved either by using spatial techniques, as low pass filtering, or frequency techniques, as the Fast Fourier Transform (FFT). On the other hand, statistical and morphological filters are typical examples of non-linear filtering. However, the filtering techniques lead in some cases to baneful effects when applied indiscriminately to an image. In fact, if it is not the whole image that is blurred, some of its important features (e.g. edges) are.

A solution to overcome this problem has been introduced by Denoho and Johnstone (Donoho & Johnstone, 1994). Instead of exploiting either linear or non-linear filtering, their technique consists of using the DWT followed by a thresholding operation. This method exploits the energy compaction ability of the wavelet transform to separate the image from the added noise. The role of the threshold is to eliminate the noise present in the image. Finally, the enhanced "denoised" image is recovered by applying the inverse DWT. This method is also known as the wavelet shrinkage denoising, and is classified as a nonlinear processing technique due to the thresholding operation involved in the process as illustrated in Figure 22.

Fig. 22. Wavelet-based denoising system

Another method, which achieves better performances when compared to the previous one, consists of using an undecimated version of the DWT (Donoho & Johnstone, 1995) This choice is motivated by the fact that originally, the DWT is not a shift-invariant transform, and as such, visual artifacts can be spanned by the transform. This like-noise is more apparent around discontinuities in the image. However, in this particular case the inverse transform is not unique. As a solution, it is appropriate to take the average of the possible reconstruction. The computational complexity of this approach is O(nlog(n)).

9.3 Image watermarking

Image watermarking emerged in the mid 90s as a discipline, among the wide range of multidisciplinary field of data hiding, as a methodology of protecting digital images from any piracy act. It consists of embedding a watermark (a trace) within a digital image before using or publishing it. The efficiency of a watermarking method lies generally in its ability to fulfil three requirements: robustness, security and invisibility.

Watermarking techniques can be classified into two categories; spatial domain methods and transform-based methods. The wavelet-based watermarking technique falls into the latter. In (Kundur & Dimitrios, 1997, 1998 & Hernandez-Guzman et al., 2008) both the original image and the watermark are first transformed to the wavelet domain, then the resulting image pyramids are fused according to certain rules, which take into account the characteristics of the Human Visual System (HVS). The wavelet in this case facilitates a simultaneous spatial localisation and frequency spread of the watermark within the source image. It has been shown that the method is robust under compression, additive noise and filtering (Kundur & Dimitrios, 1997, 1998)

To the best of our knowledge, there is no general baseline framework for a wavelet-based watermarking system. However, in most cases, the multiresolution feature of the transform is exploited to achieve robust image watermarking implementations (Kundur & Dimitrios, 1997, 1998; Tsekeridou & Pitas, 2000; Wu et al., 2000 & Hernandez-Guzman et al., 2008).

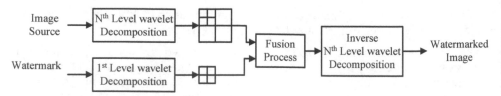

Fig. 23. Wavelet-based watermarking system

10. FPGA implementation

Quick time-to-market, low cost and high performance are typically the treble that digital system designers wish to achieve when developing new products. Although, each goal taken individually is possible, the set of three is generally beyond the capabilities of traditional design and implementation approaches (Villasenor et al., 1995; Villasenor & Mangione-Smith, 1997; Barr, 1998; Ritter & Molitor, 2000; Chrysafis & Ortega, 2000; Lafruit et al., 2000; Russel & Wayne, 2001; Ebrahimi et al., 2002; Nibouche, et al., 2000, 2001a, 2001b, 2001c, 2001d, 2002, 2003; Katona et al., 2006; Angelopoulou et al., 2008 & Lande et al., 2010). Versatile hardware such as general purpose processors (GPP), for example, can perform a wide range of operations and tasks, but fails to reach the system speed of a more specialised hardware. On the contrary, an oriented application-specific hardware, such as Application Specific Integrated Circuits (ASICs), can perform a restricted set of operations/tasks more quickly, however, at the cost of losing in generality. Hence, reconfigurable computing, generally in the form of Field Programmable Gate Arrays (FPGAs), appears to be the promising land for hardware designers. This is old/new paradigm allies the flexibility of software while preserving the hardware performances. This leads to a good trade-off between speed and generality. Unlike the case of custom hardware in the form of ASICs, which cannot be reused for a slightly different problem to the one they were designed for, configurable hardware based FPGAs allows modifications at almost any stage of the design process. In fact, configurable hardware is easily upgraded (due to its inherent nature) to suit any changes of a primal design. Used in a desktop, reconfigurable hardware can be tailored to speed up or accelerate applications, which require a system speed superior to that offered by general purpose processors. The hardware here needs to adapt itself to continual changes in response to end users needs. Obviously, the reconfigurable capabilities of such hardware will not eliminate the need for general-purpose microprocessors running on today's Personal Computers (PCs). In fact, *"FPGAs will never replace microprocessors for general-purpose computing tasks"*, as stated by Villasenor J. and Mangione-Smith W. in (Villasenor & Mangione-Smith, 1997).

The idea of reconfigurable computing was introduced first at the late 60s at the University of California at Los Angeles (UCLA) (Villasenor & Mangione-Smith, 1997 & Barr, 1998). However, the real emergence of this new paradigm for hardware computation was piloted by the commercialisation of the first SRAM-based FPGA by Xilinx Corporation in 1986 (Russel & Wayne, 2001). The first configurable devices from both Xilinx Corporation and Altera Corporation, composed typically of a fine grained structure, allowed a system speed in the range of 2MHz – 5MHz and a chip area of less than a 100 of logic blocks (Russel & Wayne, 2001). The efforts deployed by academicians and industrials since then brought to light new developments but also new challenges. In fact, the reconfigurable hardware field has dramatically maturated either by the developments in the microelectronic technology, which led to the emergence of a new range of devices providing a system gate beyond a million (e.g. Xilinx Virtex family) or by the continual emergence of a wide range of FPGA based system.

In general, FPGA devices are organised as 2D arrays of configurable logic blocks or logic elements. The parallel nature of FPGA devices make them very good targets for application that require parallel processing such as in image and video processing. In such applications, these FPGA devices are used either as co-processors or accelerators (real time applications). It is not the aim of this section to survey the field of wavelet based FPGA implementation but rather to highlight some implementation of the DWT for application in the field of image/video processing (in line with section 9).

Due to its high computational complexity, real time video compression has always been a very challenging topic for digital system designers. The implementation of such systems on FPGAs does not fail to the rule. In probably one of the earliest works in the field, Villasenor et al. in (Villasenor et al., 1995) investigated wavelet transforms based video compression algorithms for use in low-power wireless communications. Using this previous work as a basis, the same authors have further described two implementations using a single FPGA (Schoner et al., 1995). In the first approach, the proposed video compression scheme is directed towards low-complexity implementations using a single in system reprogrammable FPGA. The optimisation of the algorithm to fit the system results in an efficient implementation, however, the system is limited to only a single compression algorithm. In the second approach, to allow more flexibility, the FPGA chip is combined with an external special purpose Video Signal Processor (VSP). The FPGA/VSP combination allows the implementation of four common compression algorithms and their execution in real time. The proposed design schemes were both implemented on a Xilinx FPGA. The first design runs at 20 frames per seconds (fps) when processing a 256x256 frames with a spacial precision of 8-bits. It includes a wavelet transform, a simplified quantiser and a run-length encoder. The second scheme is capable of implementing a DCT, a 2-D FIR, a Vector Quantisation scheme (VQ) and the wavelet transform using a single generic equation. It delivers different performances: 13.3 fps for 7x7 mask 2-D filter, 55 fps for an 8x8 block DCT, 7.4 fps for a 4x4 VQ (at 1/2 bit per pixel) and 35.7 fps for a single wavelet stage.

Partitionning images prior to computation is a well known technique in the field of image processing. It has been widely used in DCT-based image compression schemes. In the last decade, this technique has been adopted in the wavelet-based JPEG2000 new compression standard (Ebrahimi et al., 2002). In (Ritter & Molitor, 2000), a biorthogonal Cohen-daubechies-Fauveau (CDF) 5/3 wavelet pair followed by Embedded Zerotree Encoding (EZT) technique is used in a lossy and a lossless compression schemes, respectively. Since the 5/3 pair is an integer-to-integer wavelet, a lifting scheme based architecture is used for the implementation. In the lossless compression scheme, the image is partitioned into a set of 32x32 tiles before processing. The system is then implemented onto an FPGA prototyping board. The system achieved an operating speed of 20MHz. In the second scheme, in order to avoid excessive increase of the internal memory, a rearrangement of the filtered and decimated outputs is proposed (interlocked external memory access. Because of its integer nature (integer to integer), as well as, for its adoption in the JPEG 2000 standard, the biorthogonal 5/3 wavelet is the focus of many studies. Since the wavelet transform algorithms are inherently multi levels, requiring complex computation schedule in hardware, a comparison of different computation schedule algorithms is presented in (Angelopoulou et al., 2008). The most widely used schedule algorithms such as the row column based algorithm (Mallat, 1989), the line based algorithm (Chrysafis & Ortega, 2000) and the block based algorithm (Lafruit et al., 2000) are implemented in FPGA using the lifting scheme and 2D DWT architecture. The 2D DWT FPGA implementation is fully parameterised. Based on the lifting scheme, Lande et al. in (Lande et al., 2010) introduce a robust invisible watermarking method to be used with still images. The scheme is incorporated in the JPEG 2000 lossless algorithm, featuring an integer to integer biorthogonal 5/3 CDF wavelet filters. The proposed algorithm targets the consumer electronics market. The objectives of the proposed FPGA implementation of this wavelet based watermarking scheme include low power usage, real time performance, robustness and ease of integration.

Denoising still images and video sequences is another field of predilection of the wavelet transform (see section 9). Katona et al. (Katona et al., 2006) suggest a real time wavelet based video denoising system and its implementation in FPGA. The method adopts a parallel approach to implement an advanced wavelet domain noise filtering algorithm, which uses a non-decimated wavelet transform. The approach relies on the wavelet "a trous" algorithm and the Daubechies minimum phase wavelet (Daub4). The proposed implementation is decentralised and distributed over two FPGAs. As a proof of concept, digitised television signals are adopted as real time video sources.

11. Conclusion

Since the late 80s, the wavelet transform has been widely used in different scientific applications including signal and image processing. This ongoing growing success, which has been characterised by the adoption of some wavelet-based schemes, is due to features inherent to the transform, such as time-scale localisation and multiresolution capabilities. In this chapter, the basic concepts of the wavelet transform have been introduced. First, the historical development of the wavelet transform and its advent to the field of signal and image processing were reviewed. Then, its features and the mathematical foundations behind it were reviewed. To ease the understanding of the wavelet theory, the related notations and terms, such as the scaling function, multiresolution, filter bank and others were described and then briefly explained.

Depending on the application at hand, different algorithms for implementing the wavelet transform have been developed. Four of these algorithms, namely, Burt's pyramid, Mallat algorithm, Feauveau's scheme and the lifting scheme were briefly described. Finally, some wavelet based image processing applications were also given.

12. References

Adelson, E. H.; Simoncelli E. & Hingorani, R. (1987). Orthogonal pyramid transforms for image coding, *SPIE Visual Communication and Image Processing II*, Vol. 845, pp. 50-58
Angelopoulou, M. E.; Cheung, P. Y. K ; Masselos, K. & Andreopoulos, Y. (2008). Implementation and comparison of the 5/3 lifting 2D discrete wavelet transform

computation schedule on FPGAs, *Journal of signal processing systems*, Vol. 51, pp. 3 - 21

Barr, M. (1998). A Reconfigurable Computing Primer, *Multimedia Systems Design*, pp. 44-47

Bradley, J.; Brislawn, C. & Hopper, T. (1993). *The FBI Wavelet/Scalar Quantization Standard for Gray-scale Fingerprint Image Compression*, Tech. Report LA-UR-93-1659, Los Alamos Nat'l Lab, Los Alamos

Brislawn, C. M. (April 2010). Group Lifting Structures for Multirate Filter Banks II: Linear Phase Filter Banks, *IEEE Transactions on Signal Processing*, Vol. 58, No. 4, pp. 2078 - 2087, ISSN 1053-587X

Burt, P. J. & Adelson, A. E. (1983). The Laplacian pyramid as a compact image code, *IEEE Transactions on Communications*, Vol. 31, No. 4, (Apr 1983), pp. 532-540, ISSN 0090-6778

Burrus, C. S.; Gopinath, R. A. & Guo, H. (1998). *Introduction to Wavelets and Wavelet Transforms: A primer*, Prentice Hall

Chappelier, V. & Guillemot, C. (2006). Oriented Wavelet Transform for Image Compression and *Denoising, IEEE Transactions on Image Processing*, Vol. 15, No. 10, pp. 2892-2903, ISSN 1057-7149

Chen, G. & Qian, S. (2011). Denoising of Hyperspectral Imagery Using Principal Component Analysis and Wavelet Shrinkage, *Geoscience and Remote Sensing, IEEE Transactions*, Vol. 49, No. 3, pp. 973 - 980, ISSN 0196-2892

Chrysafis, C. & Ortega, A. (2000). Line based reduced memory wavelet image compression, *IEEE Transactions on Image Processing*, Vol. 9, No. 3, pp. 378-389, 010-1024, ISSN 1057-7149

Cunha, A. L.; Zhou, J. & Do, M. N. (October 2006). The nonsubsampled contourlet transform: Theory, design, and applications, *IEEE Transactions on Image Processing*, Vol. 15, No. 10, pp. 3089–3101, ISSN: 1057-7149

Cohen, A.; Daubechies, I. & Feauveau, J. (1992). Biorthogonal bases of compactly supported wavelets, *Communications on Pure and Applied Mathematics*, Vol. 45, No. 5, pp. 485-560

Cody, M. A. (1994). The Wavelet Packet Transform, *Dr. Dobb's Journal*, Vol. 19, Apr. 1994

Daubechies, I. (1988). Orthonormal bases of compactly supported wavelets, *Communications on Pure and Applied Mathematics*, Vol. 41, pp. 909-996

Daubechies, I. (1992). Ten lectures on Wavelets, *SIAM*, Philadelphia

Daubechies, I. (Mar. 1993). Orthonormal bases of compactly supported wavelets II, variations on a theme, *SIAM Journal of Mathematical Analysis*, Vol. 24, No. 2, pp. 499-519

David F. W. (2002). *Wavelet Analysis*, Birkhauser, ISBN-0-8176-3962-4

Do, M. N. & Vetterli, M. (January 2003). The finite ridgelet transform for image representation. *IEEE Transactions on Image Processing*, Vol. 12, No. 1, pp. 6–28, ISSN 1057-7149

Do, M. N. & Vetterli, M. (December 2005). The contourlet transform: An efficient directionalmultiresolution image representation. *Transactions on Image Processing*, Vol. 14, No. 12, pp. 2091-2106, ISSN 1057-7149

Donoho, D. L. & Johnstone, I. M. (1994). Ideal Spatial Adaptation via Wavelet Shrinkag, *Biometrika*, Vol. 81, No. 3, pp. 425-455, Online ISSN 1464-3510 - Print ISSN 0006-3444

Donoho, D. L. & Johnstone, I. M. (Dec. 1995). Adaptation to Unknown Smoothness Via Wavelet Shrinkage, *Journal of American Statistical Association*, Vol. 90, No. 432, pp. 1200-1224, ISSN 01621459

Ebrahimi, T.; Christopoulos, C. & Lee, D. L. (Eds) (Jan. 2002). JPEG2000, *Special Issue of Signal Processing: Image Communication*, Vol. 17, No. 1, Elsevier Science

Feauveau, J. C. (1990). *Analyse multiresolution par ondelettes non-orthogonales et bancs de filtres numeriques*. PhD Thesis, Universite Paris Sud, France

Firoiu, I.; Nafornita, C.; Boucher, J. M. & Isar, A. (2009). Image Denoising Using a New Implementation of the Hyperanalytic Wavelet Transform, *IEEE Transactions on Instrumentation and Measurement*, Vol. 58, No. 8, pp. 2410 – 2416, ISSN 0018-9456

Fliege, N. J. (1994). *Multirate Digital Signal Processing: Multirate Systems, Filter Banks, Wavelets*, John Wiley & Sons, ISBN: 0471939765, Inc. New York, NY, USA

Grossman, A. & Morlet, J. (1984). Decomposition of hardy functions into square integrable wavelets of constant shape, *SIAM Journal of Mathematical Analysis*, Vol. 15, No. 4, pp. 723-736, ISSN 00361410

Graps, A. L. (1995). An introduction to Wavelets, *IEEE Computational Science and Engineering*, Vol. 2, No. 2, pp. 50-61, ISSN: 1070-9924

Hankerson, D. C.; Harris, G. & Johnson, P. (2005). *Introduction to Information Theory and Data Compression*, Taylor & Francis e-Library, ISBN 1-58488-313-8

Hernandez-Guzman, V. ; Cruz-Ramos, C. ; Nakano-Miyatake M. & Perez-Meana, H. (2008). Watermarking Algorithm based on the DWT, *Latin America Transactions, IEEE (Revista IEEE America Latina)*, Vol. 4, No. 4, pp. 257-267, ISSN 1548-0992

Katona, M.; Pizurica, A.; Teslic, N.; Kovacevic, V. & Philips, W. (2006). A Real time wavelet domain video denoising implementation in FPGA, *EURASIP Journal of Embedded Systems, Hindawi Publishing*, Vol. 2006, No. 1, pp. 1-12, ISSN 1687-3955, EISSN 1687-3963

Kundur D. & Dimitrios H. (1997). A robust digital image watermarking method using wavelet-based fusion (1997), *ICIP (1)'1997, Proc. of IEEE Int. Conf. on Acoustics, Speech and Sig. Proc., vol. 5, pp. 544-547 Seattle,Washington (1997-5)*

Kundur D. & Dimitrios H. (1998). Digital watermarking using multiresolution wavelet decomposition, *In Proceedings of IEEE ICAPSSP '98*, Vol. 5, pp. 2969 – 2972, Seattle, WA, USA, May 1998

Lafruit, G.; Nachtergaele, L.; Vanhoof, B. & Catthoor, F. (2000). The Local Wavelet Transform: a Memory Efficient, High Speed Architecture Optimised to Region Oriented Zero Tree Coder, *Integrated Computer Aided Engineering Journal*, Vol. 7, No. 2, pp. 89-103, ISSN:1069-2509

Lande, P. U. ; Talbar, S. N. & Shinde, G. N. (2010). FPGA Prototype of Robust Image Watermarking For JPEG 2000 With Dual Detection, *Int. Journal of Computer Science and Security*, Vol. 4, No. 2, pp 226-236, ISSN 1985-1553

Lu, Y. & Do, M. N. (April 2007). Multidimensional Directional Filter Banks and Surfacelets, *In Image Processing, IEEE Transactions*, Vol. 16, No. 4, pp. 918-931, ISSN: 1057-7149

Mallat, S. (July 1989). A theory for multiresolution signal decomposition: the wavelet representation, *IEEE Transactions on Pattern Recognition and Machine Intelligence*, Vol. 11, No. 7, pp. 674-693, ISSN 0162-8828

Mallat, S. (2009). *A Wavelet Tour of Signal Processing, Third Ed…(Hardcover)*,Copyrighted Material, Elsevier Inc. ISBN 13: 978-0-12-374370-1

A theory for multiresolution signal decomposition: the wavelet representation, *IEEE Transactions on Pattern Recognition and Machine Intelligence*, Vol. 11, No. 7, pp. 674-693, ISSN 0162-8828

Masud, S. (1999). *VLSI system for Discrete Wavelet Transforms*, PhD thesis, Department of Electrical Engineering, The Queen's University of Belfast, Ireland

Meyer, Y. (1987). Wavelet with compact support. *Zygmund Lectures*, University Chicago 1987

Nai-Xiang L. & Yap-Peng T. (2005). Color image denoising using wavelets and minimum cut analysis, *Signal Processing Letters IEEE*, Vol. 12, No. 11, pp. 741-744, ISSN 1070-9908

Nai-Xiang, L.; Vitali, Z. & Yap-Peng T. (2006). Error inhomogeneity of wavelet image compression, *IEEE*, pp. 1597-1600, ISSN 1522-4880

Nguyen, T. T. & Oraintara, S. (Oct. 2008). The Shiftable Complex Directional Pyramid — Part I: Theoretical Aspects, *IEEE Transactions on Signal Processing*, Vol. 56, No. 10, pp. 4651-4660, ISSN 1053-587X

Nibouche, M.; Bouridane, A.; Nibouche, O.; Crookes, D. & Boussekta, S. (2000). Design and FPGA implementation of orthonormal discrete wavelet transforms, *The 7th IEEE International Conference, Electronics, Circuits and Systems, ICECS 2000*, Vol. 1, pp. 312-315, ISBN 0-7803-6542-9

Nibouche, M.; Bouridane, A.; Crookes, D. & Nibouche, O. (2001a). An FPGA-based wavelet transforms coprocessor, *International Conference on Image Processing, Proceedings 2001*, Vol. 3, pp. 194-197, ISBN: 0-7803-6725-1

Nibouche, M.; Bouridane, A.; Nibouche, O. & Crookes, D. (2001b). Rapid prototyping of orthonormal wavelet transforms on FPGAs, *Circuits and Systems, 2001. ISCAS 2001. The 2001 IEEE International Symposium*, Vol. 2, pp. 577-580, ISBN 0-7803-6685-9

Nibouche, M.; Bouridane, A.; Nibouche, O. & Belatreche, A. (2001c). Design and FPGA implementation of orthonormal inverse discrete wavelet transforms, *Wireless Communications, 2001. (SPAWC '01). 2001 IEEE Third Workshop on Signal Processing Advances*, pp. 365-359, ISBN 0-7803-6720-0

Nibouche, M.; Bouridane, A. & Nibouche, O. (2001d). A framework for a wavelet-based high level environment, *Electronics, Circuits and Systems, 2001. ICECS 2001. The 8th IEEE International Conference*, Vol. 1, pp. 429-432, ISBN 0-7803-7057-0

Nibouche, M. & Nibouche, O. (2002). Design and implementation of a wavelet block for signal processing applications, *Electronics, Circuits and Systems, 2002. 9th International Conference*, Vol.3, pp. 867-870, ISBN 0-7803-7596-3

Nibouche, M.; Nibouche, O. & Bouridane, A. (Dec. 2003). Design and implementation of a wavelet based system, *Electronics, Circuits and Systems, 2003. ICECS 2003. Proceedings of the 2003 10th IEEE International Conference*, Vol. 2, pp. 463-466, ISBN 0-7803-8163-7

Oppenheim, A. V. & Schafer, R. W. (2010). *Discrete-Time Signal Processing*, Prentice Hall, Upper Saddle River, NJ, third edition, ISBN-13 978-0613-198842-2/ISBN-10 0-13-198842-5

Raviraj, P. & Sanavullah, M. Y. (Apr-Jun 2007). The modified 2D-Haar Wavelet Transformation in image compression, *Middle East Journal of Scientific Research*, Vol. 2 , No .2, pp. 73-78, ISSN 1990-9233

Ritter, J. & Molitor, P. (2000). A Partitioned Wavelet-based Approach for Image Compression using FPGAs, *Proceedings of the IEEE Custom Integrated Circuits Conference (CICC)*, pp. 547-550, Orlando, Florida, USA, May 2000

Ruikar, S. & Doye, D. D. (2010). Image denoising using wavelet transform. *IEE Mechanical and Electrical Technology (ICMET), 2010 2nd International Conference*, pp. 509-515, ISBN 978-1-4244-8100-2

Russel T. & Wayne B. (2001). Reconfigurable Computing for Digital Signal Processing: A Survey, *Journal of VLSI Signal Processing systems*, Vol. 28, pp 7-27, ISSN 0922-5773

Schoner, B.; Villasenor, J.; Molloy, S. & Jain, R. (1995). Techniques for FPGA Implementation of Video Compression Systems, *Proceedings of the Third International ACM Symposium on FPGA '95.*, pp. 154-159, ISBN: 0-7695-2550-4

Smith, S. (2003). *Digital Signal Processing : A practical guide for engineers and scientists*, Elsevier, ISBN 0-75067444-X, USA

Starck, J. L.; Murtagh, F. & Bijaoui, A. (1998). Image Processing and Data Analysis: The multiscale approach, *Cambridge University Press*, 1998, First published 1998, Reprinted 2000, ISBN 0521 59084 1- ISBN 0521 59914 8, U. K.

Strang G. & Nguyen T. (1996). Wavelets and Filter Banks, *Wellesley-Cambridge Press*, ISBN 09614088-7-1, USA

Stromme, O. (1999). *On the applicability of wavelet transforms to image and video compression*, PhD thesis, Department of Computer Science, the University of StrathClyde

Sweldens, W. (1996). Wavelets: what next?, *Proceedings of the IEEE*, vol. 84, No. 4, pp. 680-685, ISSN 0018-9219

Sweldens, W. (1995). The Lifting Scheme: A new philosophy in biorthogonal wavelet constructions, *Wavelet Applications in Signal and Image Processing III*, pp. 68-79, Proc. SPIE 2569

Tsekeridou, S. & Pitas, I. (2000). Wavelet-based Self-similar Watermarking for Still Images, *IEEE International Symposium on Circuits and Systems*, Geneva, Switzerland, Print ISBN 0-7803-5482-6

Valens, C. (2004). A Really Friendly Guide To Wavelets, Latest version: V02122004, available from http://perso.wanadoo.fr/polyvalens/clemens/

Villasenor, J; Jones, C. & Schoner, B. (1995). Algorithms and System Prototype for Low-Power, Low-bit-rate Wireless Video Coding, *IEEE Transactions on Circuits and Systems for Video Technology*, Vol.5, No. 6, pp 565-567, ISSN 1051-8215

Villasenor, J. and Mangione-Smith, W. (1997). Configurable Computing, *Scientific American*, June 1997

Wickerhauser, M. D. (1994). *Adapted Wavelet Analysis from Theory to Software*, First Edition, ISBN 1568810415/1-56881-041-5, A. K. Peters Ltd, Natick, Massachusetts, U.S.A

Wu, X.; Zhu, W.; Xiong Z. & Zhang Y. (2000). Object-based Multiresolution Watermarking of Images and Video, *IEEE International Symposium on Circuits and Systems*, vol. 1, pp. 545–550, Geneva, Switzerland

Wavelet Based Image Compression Techniques

Pooneh Bagheri Zadeh[1], Akbar Sheikh Akbari[2] and Tom Buggy[2]
[1]Staffordshire University,
[2]Glasgow Caledonian University
UK

1. Introduction

With advances in multimedia technologies, demand for transmission and storage of voluminous multimedia data has dramatically increased and, as a consequence, data compression is now essential in reducing the amount of data prior storage or transmission. Compression techniques aim to minimise the number of bits required to represent image data while maintaining an acceptable visual quality. Image compression is achieved by exploiting the spatial and perceptual redundancies present in image data. Image compression techniques are classified into two categories, lossless and lossy. Lossless techniques refer to those that allow recovery of the original input data from its compressed representation without any loss of information, i.e. after decoding, an identical copy of the original data can be restored. Lossy techniques offer higher compression ratios but it is impossible to recover the original data from its compressed data, as some of the input information is lost during the lossy compression. These techniques are designed to minimise the amount of distortion introduced into the image data at certain compression ratios. Compression is usually achieved by transforming the image data into another domain, e.g. frequency or wavelet domains, and then quantizing and losslessy encoding the transformed coefficients (Ghanbari, 1999; Peng & Kieffer, 2004; Wang et al., 2001). In recent years much research has been undertaken to develop efficient image compression techniques. This research has led to the development of two standard image compression techniques called: JPEG and JPEG2000 (JPEG, 1994; JPEG 2000, 2000), and many non-standard image compression algorithms (Said & Pearlman, 1996; Scargall & Dlay, 2000; Shapiro, 1993).

Statistical parameters of image data have been used in a number of image compression techniques (Chang & Chen, 1993; Lu et al., 2000; Lu et al., 2002; Saryazdi and Jafari, 2002). These techniques offer promising visual quality at low bit rates. However, the application of statistical parameters of the transformed data in image compression techniques has been less reported in the literature. Therefore, the statistical parameters of the transformed image data and their application in developing novel compression algorithms are further investigated in this chapter.

The performance of image compression techniques can also be significantly improved by embedding the properties of the Human Visual System (HVS) in their compression algorithms (Bradley, 1999; Nadenau et al., 2003). Due to the space–frequency localization properties of wavelet transforms, wavelet based image codecs are most suitable for embedding the HVS model in their coding algorithm (Bradley, 1999). The HVS model can be embedded either in the quantization stage (Aili et al., 2006; HSontsch & Karam, 2000; Nadenau et al., 2003), or at the bit allocation stage (Antonini et al., 1992; Sheikh Akbari & Soraghan, 2003; Thornton et al., 2002; Voukelatos & Soraghan, 1997) of the wavelet based encoders. In this chapter, HVS coefficients for wavelet high frequency subbands are calculated and their application in improving the coding performance of the statistical encoder is investigated.

2. Fundamental of compression

The main goal of all image compression techniques is to minimize the number of bits required to represent a digital image, while preserving an acceptable level of image quality. Image data are amendable to compression due to the spatial redundancies they exhibit and also because they contain information that, from a perceptual point of view, can be considered irrelevant. Many standard and non-standard image compression techniques have been developed to compress digital images. These techniques exploit some or all of these image properties to improve the quality of the decoded images at higher compression ratios. Some of these image coding schemes are tabulated in Table 1.

Image compression techniques can be classified into two main groups, named: lossless and lossy compression techniques. In lossless compression process, the original data and the reconstructed data must be identical for each and every data sample. Lossless compression is demanded in different applications such as: medical imagery, i.e. cardiography, to avoid the loss of data and errors introduced into the imagery. Also, it is applied to the case that is not possible to determine the acceptable loss of data.

In most image processing applications, there is no need for the reconstructed data to be identical in value with its original. Therefore, some amount of loss is permitted in the reconstructed data. This kind of compression techniques, which results in an imperfect reconstruction, is called lossy compression. By using lossy compression, it is possible to represent the image with some loss using fewer bits in comparison to a lossless compression.

3. Characteristics of the Human Visual System

Research has shown that embedding the Human Visual System (HVS) model into compression algorithms yields significant improvement in the visual quality of the reconstructed images (Aili et al., 2006; Antonini et al., 1992; Bradley, 1999; HSontsch & Karam, 2000; Nadenau et al., 2003; Sheikh Akbari & Soraghan, 2003; Thornton et al., 2002; Voukelatos & Soraghan, 1997). It has been shown in (Bradley, 1999; Nadenau et al., 2003) that the performance of image compression techniques can be significantly improved by exploiting the limitations of the HVS for compression purposes. To achieve this aim, the HVS-model can be embedded in the compression algorithm to optimise the perceived visual quality.

Standard image coding techniques		Non standard image coding techniques
DCT-base	Wavelet-base	Differential Pulse code Modulation (DPCM) Vector Quantization (VQ) Zero-Tree Coding Fractal Neural Networks Trellis Coding
JPEG (1980)	JPEG2000 (2000)	

Table 1. Standard and non-standard image compression techniques.

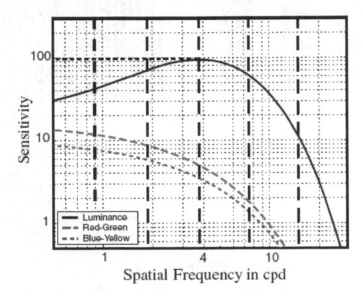

Fig. 1. The CSF curves for the luminance and chrominance channels of the HVS (Nadenau et al., 2003).

Due to the complexity of the human visual processing system, assessments of the performance of HVS-models are based on psychophysical observations. Physiologists have performed many psycho-visual experiments with the goal of understanding how the HVS works. One of the limitations of the HVS, which was found experimentally, is the lower sensitivity of the HVS for patterns with high spatial-frequencies. Exploiting this property of the HVS model, and embedding it into compression algorithms, can significantly improve the visual quality of compressed images. Natural images are composed of small details and shaped regions. Therefore, it is necessary to describe the contrast sensitivity as a function of spatial frequency. This phenomenon has been known as the Contrast Sensitivity Function (CSF) (Nadenau et al, 2003; Tan et al, 2004). Figure 1 shows the CSF curves for the luminance and chrominance channels of the HVS. From Figure 1, it can be seen that the HVS is more sensitive to the luminance component than the chrominance components. The sensitivity of the HVS in terms of luminance is greatest around the mid-frequencies, in the region of 4 cycles per optical degree (cpd). It rapidly reduces at higher spatial frequencies, and slightly decreases at lower frequencies. The HVS, in terms of chrominance components behaves like a low pass-filter; therefore there is no decrease in its sensitivity at low frequencies.

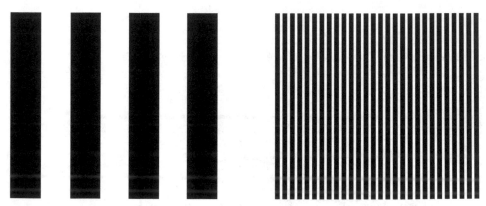

Fig. 2. A low frequency pattern (left) and a high frequency pattern (right), the high frequency pattern appear less intense.

To give a sense of the sensitivity of HVS to different frequencies, two black and white patterns are shown in Figure 2, a low frequency pattern on the left and a high frequency pattern on the right. In both patterns, the black and white have the same brightness, but the black and white colours of the right hand pattern appears less intense than the pattern in the left side. This can be explained by the fact that the HVS is less sensitive to high frequency components.

3.1 Human Visual System in compression techniques

Wavelet-based image coding schemes have proven to be ideally suited for embedding complete HVS models, due to the space–frequency localization properties of the wavelet decompositions (Bradley, 1999). The HVS model has been embedded either at the quantization stage (Aili et al., 2006; HSontsch & Karam, 2000; Nadenau et al., 2003), or at the bit allocation stage (Antonini et al., 1992; Sheikh Akbari & Soraghan, 2003; Thornton et al., 2002; Voukelatos & Soraghan, 1997) of the codec, which yields significant improvement in the visual quality of the reconstructed images. Antonini et al. (Antonini et al., 1992) developed a wavelet-based image compression scheme using Vector Quantization (VQ) and the property of the HVS. This algorithm performs a Discrete Wavelet Transform (DWT) on the input image and then the resulting coefficients in different subbands are vector quantized. The bit allocation among different subbands is based on a weighted Mean

Squared Error (MSE) distortion criterion, where the weights are determined based on the property of the HVS introduced in (Campbell & Robson, 1968). (Voukelatos & Soraghan, 1998) introduced another wavelet based image compression technique using VQ and the properties of the HVS. They first calculated the value of the Contrast Sensitivity Function (CSF) for the central spatial frequency of each subband. These values were then used to scale the threshold value for each subband, which were used in vector selection prior to the VQ process. A weighted MSE distortion criterion using perceptual weights is also employed to allocate bits among different subbands. Voukelatos and Soraghan reported significant improvement over existing block-based image compression techniques at very low bitrates. Thornton et al (Thornton et al., 2002) extended the Voukelatos and Soraghan's algorithm (Voukelatos & Soraghan, 1998) to video for very low bitrate transmission. Thornton et al. incorporated the properties of the HVS to code the intra-frames and reported significant improvement in objective visual quality of the decompressed video sequences. Sheikh Akbari and Soraghan (Sheikh Akbari & Soraghan, 2003) developed another wavelet based video compression scheme using the VQ scheme and the properties of the HVS. They calculated the value of the CSF for the central spatial frequency of each subband of the Quarter Common Intermediate (QCIF) image size. These values were then used to scale the threshold value for each subband, which were used in vector selection prior to the VQ process and also in the bit allocation among different subbands.

The JPEG 2000 standard image codec supports two types of visual frequency weighting: Fixed Visual Weighting (FVW) and Visual Progressive Coding or Visual Progressive Weighting (VPW). In FVW, only one set of CSF weights is chosen and applied in accordance with the viewing conditions, and in the VPW, different sets of CSF weights are used at the various stages of the embedded coding. This is because during a progressive transmission stage, the image is viewed at various distances. For example, at low bitrates, the image is viewed from a relatively large distance, while as more bits are received, the quality of the reconstructed image is increased, which implies that the viewer looks at the image from a closer distance (Skodras et al, 2001). Nadenau et al. incorporated the characteristic of the HVS into a wavelet-based image compression algorithm using a noise-shape filtering stage prior to the quantization stage (Nadenau et al., 2003). They filtered the transformed coefficients using a "HVS filter" for each subband. This algorithm improves the compression ratio up to 30% over the JPEG2000 baseline for a number of test images. A new image compression method based on the HVS was proposed by Aili et al. (Aili et al., 2006). In this codec, the input image is first decomposed using a wavelet transform, and then the transformed coefficients in different subbands are weighted by the peak of the contrast sensitivity function (CSF) curve in the wavelet domain. Finally the weighted wavelet coefficients were coded using the Set Partitioning in Heretical Tree (SPIHT) algorithm. This technique showed significantly higher visual and almost the same objective quality to that of the conventional SPIHT technique.

3.2 Calculation of perceptual weights

In this section, the perceptual weights that regulate the quantization steps in different image compression techniques are specifically calculated for a Quarter Common Intermediate Format (QCIF) image size. The derivation of the weighting factors is based on the results of subjective experimental data that was presented in (Van Dyck & Rajala, 1994).

3.2.1 Calculation of spatial frequencies

The perceptual coding model is designed for a QCIF image size, thus this corresponds to a physical dimension of 1.8×2.2 inches on the workstation monitor, i.e. videophone display. Therefore, the pixel resolution r, which is measured in pixels per inch, in both the horizontal and vertical dimensions, will be 80 pixels/inch. Let us assume the viewing distance v, which is measured in metres, be 0.30 metres. This distance is a good approximation of the natural viewing distance of a human using a videophone device. The sampling frequency, f_s in pixels per degree, can be then calculated using Equation 1 (Nadenau et al., 2003):

$$f_S = \frac{2\,v\tan(0.5^\circ)r}{0.0254} \tag{1}$$

The signal is critically down-sampled at Nyquist rate to 0.5 cycle/pixel. Hence the maximum frequency represented in the signal is:

$$f_{max} = 0.5 f_S \tag{2}$$

Thus the maximum frequency represented in the QCIF image size with the thirty centimetre distance will be 8.246 cycles/degree. The centre radial frequency for each subband is determined by the Euclidean distance of its centre from the origin where subbands are in a square of length 8.246 and the base-band is in the origin. Figure 3 shows the centre radial frequencies for each sub-band of a three level wavelet decomposition.

3.2.2 Mean detection threshold

The mean detection threshold is the smallest change in a colour that is noticeable by a human observer and is used to calculate the perceptual weighting factors. It is a function of spatial frequency, orientation, luminance and background colour. The initial data presented in (Van Dyck & Rajala, 1994) was measured in the xyY colour space, where x and y are the C.I.E. chromaticity coordinates and Y is the luminance. Table 2 gives the set of thresholds for various frequencies and orientations measured along the luminance, Red-Green and Blue-Yellow directions when the luminance value Y_0 is 5 cd / m^2 and background colour is white. The chromaticity coordinates for white are: $(x_0, y_0) = (0.33, 0.35)$. For transition along the Red-Green and Blue-Yellow direction each mean detection threshold gives two chromaticity coordinates corresponding to the maximum and minimum of the sinusoidal variation as shown in equations 3 and 4, respectively.

$$x_i = x_0 \pm \Delta x \cdot t \tag{3}$$

$$y_i = y_0 \pm \Delta y \cdot t \tag{4}$$

where t is the mean detection threshold, Δx and Δy are the step sizes for the changes in the x and y direction. The values used for Δx and Δy for all three directions are given in Table 3.

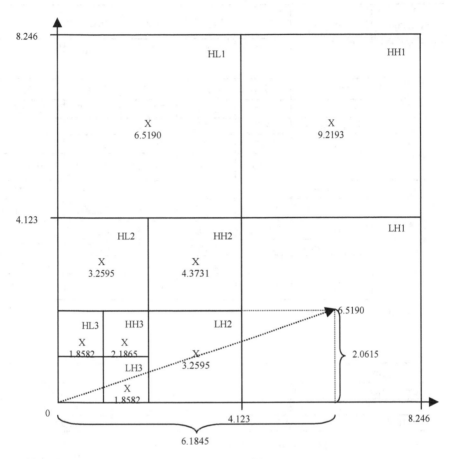

Fig. 3. Subband centre spatial frequencies in cycles/degree.

3.2.3 Perceptual weight factors

The perceptual weight for each subband is the reciprocal of its mean detection threshold. Hence, the mean detection thresholds for the YIQ space need to be calculated before the perceptual weights can be determined. The mean detection thresholds in the xyY space for the centre frequencies of the subbands shown in Figure 3 are first calculated by linearly interpolating the values in Table 2. In wavelet decomposition, the diagonal subbands (HH) do not discriminate between left and right, so an average of the two values is employed. The resulting thresholds in the xyY space for the centre of the high frequency subbands are listed in Table 4. By using equations 3 and 4, two chromaticity coordinates $(x_i$, y_i , $Y_0)$, where $i = 1$, 2 for each subband can be calculated. These two chromaticity coordinates are in the xyY space. Therefore they are converted from the xyY space to the C.I.E. XYZ space using the equations in 5: (Ghanbari, 1999).

Spatial Direction	Colour Direction	Spatial frequency cycles/deg				
		1.0	2.0	5.0	10.0	20.0
Horizontal (LH)	Luminance	6.750	6.330	7.250	13.500	65.083
	R-G	4.750	4.750	7.617	17.417	77.417
	B-Y	6.000	6.833	32.667	70.167	150.000
Vertical (HL)	Luminance	6.833	6.250	6.833	22.500	77.800
	R-G	5.583	7.083	9.250	23.000	90.375
	B-Y	6.667	9.417	31.833	65.700	150.000
Left Diagonal (HH)	Luminance	7.667	6.917	11.167	37.083	49.000
	R-G	7.917	7.167	16.083	37.500	100.750
	B-Y	12.417	18.500	45.500	86.500	150.000
Right Diagonal (HH)	Luminance	8.083	7.583	9.167	42.583	85.750
	R-G	7.750	6.333	13.833	35.417	103.500
	B-Y	13.750	19.750	47.750	83.000	114.000

Table 2. Mean detection thresholds in xyY space (Van Dyck 1994).

Direction	ΔY	Δx	Δy
Luminance	0.0124	0.0	0.0
R-G	0.0	0.000655	-0.000357
B-Y	0.0	0.000283	0.000689

Table 3. Step size for changes in each direction (Van Dyck 1994).

SUBBAND	Mean Detection Threshold		
	Luminance	R-G	B-Y
LH1	8.731	9.939	41.554
HL1	10.546	12.508	39.859
HH1	32.436	48.890	75.188
LH2	6.664	5.793	16.236
HL2	6.462	7.871	17.576
HH2	9.556	13.242	40.877
LH3	6.520	4.750	6.454
HL3	6.514	6.402	8.168
HH3	7.431	7.261	20.839

Table 4. Mean detection thresholds in xyY space for subbands.

$$X_i = x_i Y_0 / y_i$$

$$Z_i = \left(1 - x_i - y_i\right) Y_0 / y_i \tag{5}$$

$$for \quad i = 1, 2$$

For the luminance direction each mean detection threshold also provides two XYZ values that are calculated using the equations in 6:

$$X_i = X_0 \pm \Delta Y. \frac{X_0}{Y_0}.t$$

$$Y_i = Y_0 \pm \Delta Y.t \tag{6}$$

$$Z_i = Z_0 \pm \Delta Y. \frac{Z_0}{Y_0}.t$$

where ΔY is given in Table 3, t is the mean detection threshold and $i = 1,2$. The vector (X_0, Y_0, Z_0) contains the coordinates of the white point, computed from equation 6. The resulting values are then transformed into the YIQ space. The Red-Green line lies approximately in the I-direction and the Blue-Yellow line lies mostly in the Q direction. The linear transformations in equations 7 and 8 are used to give two points for each direction in the YIQ space.

$$\begin{bmatrix} R \\ G \\ B \end{bmatrix} = \begin{bmatrix} 1.910 & -0.533 & -0.288 \\ -0.985 & 2.000 & -0.028 \\ 0.058 & -0.118 & 0.896 \end{bmatrix} . \begin{bmatrix} X \\ Y \\ Z \end{bmatrix} \tag{7}$$

$$\begin{bmatrix} Y \\ I \\ Q \end{bmatrix} = \begin{bmatrix} 0.299 & 0.587 & 0.114 \\ 0.596 & -0.274 & -0.322 \\ 0.211 & -0.523 & 0.312 \end{bmatrix} . \begin{bmatrix} R \\ G \\ B \end{bmatrix} \tag{8}$$

The YIQ mean detection threshold for each direction is the inverse Euclidean distance between these two points. The computed weighting factors for each subband of QCIF video, based on the properties of the HVS, are shown in Table 5. These values represent the perceptual weights that can be used to regulate the quantization step-size in the pixel quantization of the high frequency subbands' coefficients of the Multiresolution based image/video codecs.

4. Statistical parameters in image compression

Statistical parameters of the image data have been used in a number of image compression techniques (Chang & Chen, 1993; Lu et al., 2000; Lu et al., 2002; Saryazdi & Jafari, 2002) and have demonstrated promising improvement in the quality of decompressed images, especially at medium to high compression ratios. A vector quantization based image compression algorithm was proposed by Chang and Chen (Chang & Chen, 1993). It first generates a number of sub-codebooks from the super-codebook, and then employs the statistical parameters of the upper and left neighbour vectors to decide which codebook is to be used for vector quantization. This coding scheme has been extended by Lu et al. (Lu et al., 2000) who generated two master-codebooks, one for the codewords whose variances are larger than a threshold, and another one for the remainder codewords. Lu et al. exploited the current vector's statistical parameter to decide which of these two master codebooks to use for vector quantization, and then Chang and Chen's algorithm was applied to perform vector quantization. Lu et al. (Lu et al., 2002) successfully developed other gradient-based vector quantization schemes and reported further improvement at low bit rates. In the Lu et al. proposed algorithms, one master codebook is first generated and codewords are then sorted in ascending order of their gradient values. In the first algorithm, Chang and Chen's (Chang and Chen, 1993) technique is used to perform vector quantization, with the difference that gradient parameters instead of statistical parameters are used to decide which codebook is to be used for vector quantization. In the second algorithm, the number of codebooks was increased, which resulted in further bit reduction. Another statistically-based image compression scheme was reported by (Saryazdi and Jafari, 2002). In this algorithm, the input image is divided to a number of blocks. The statistical parameters are then used to classify each block into uniform and non-uniform blocks. The uniform blocks are coded by their minimum values. The non-uniform blocks are coded by their minimum and residual values, where the residual values are vector quantized. They reported promising visual quality at high compression ratios.

SUBBAND	Y-DOMAIN	I-DOMAIN	Q-DOMAIN
LH1	4.3807	2.0482	1.0502
HL1	3.4573	1.6159	1.0992
HH1	1.2372	0.6978	0.6065
LH2	5.9673	3.6449	2.6340
HL2	6.1708	2.7149	2.4728
HH2	4.1934	1.6384	1.1331
LH3	6.1796	4.5685	7.1443
HL3	6.1984	3.3243	5.5495
HH3	5.3931	2.9888	2.2339

Table 5. Perceptual weight factors for the YIQ colour domain.

4.1 Distribution of wavelet transform coefficients

Wavelet transform is one of the most popular transform that has been used in many image-coding schemes. As each statistical distribution function has its own parameters, knowledge of the statistical behaviour of the wavelet transformed coefficients in each subband of an image, can play an important role in designing an efficient compression algorithm. Study of many non-artificial images has shown that distribution of the wavelet-transformed coefficients in high frequency subbands of natural images follow a Gaussian distribution (Altunbasak & Kamaci, 2004; Kilic & Yilmaz, 2003; Eude et al., 1994; Valade & Nicolas, 2004; Yovanof & Liu, 1996). In the following, the Gaussian distribution and its statistical parameters are first reviewed. Then, a review of the study on the distribution of the wavelet transform-coefficients of images is given. A one dimensional Gaussian distribution function $f_g(x)$ is defined as follow:

$$f_g(x) = \frac{1}{\sqrt{2\pi\sigma^2}}\, e^{\frac{-(x-\mu)^2}{2\sigma^2}} \tag{9}$$

where μ is the mean value of $f_g(x)$ and is calculated using Equation 10:

$$\mu = \int_{-\infty}^{+\infty} x\, f_g(x)\, dx \tag{10}$$

and σ is known as the standard deviation, which determines the width the of the distribution. The square of the standard deviation, σ^2, is called the variance and is determined as follows:

$$\sigma^2 = \int_{-\infty}^{+\infty} x^2\, f_g(x)\, dx \tag{11}$$

where the mean value, μ, and variance, σ^2, of discrete data, are calculated using Equations 12 and 13, respectively.

$$\mu = \frac{1}{n} \cdot \sum_{i=1}^{n} x_i \tag{12}$$

$$\sigma^2 = \frac{1}{n} \cdot \sum_{i=1}^{n} (x_i - \mu)^2 \tag{13}$$

where n is the number of the discrete data, and x_i is the data. Every Gaussian distribution function is defined by two parameters: the mean value, which defines the central location of the distribution, and the variance, which defines the width of the distribution. Four Gaussian distribution functions, with different mean values and variances, are shown in Figure 4.

Study of the distribution of wavelet transform coefficients in each subband has shown that the distribution of the coefficients in the detail subbands of the wavelet-transformed data of natural images is approximately Gaussian (coefficients in the baseband are excluded) [Valade and Nicolas, 2004][Kilic and Yilmaz, 2003]. Distributions of the wavelet coefficients of an image, after applying a three level 2D-wavelet transform, are shown in Figure 5. From Figure 5, it can be seen that except for the lowest frequency coefficients, the distribution of the coefficients in high frequency subbands is approximately Gaussian.

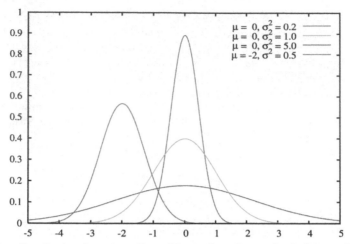

Fig. 4. Gaussian distribution functions (http://en.wikipedia.org/wiki/Normal - distribution).

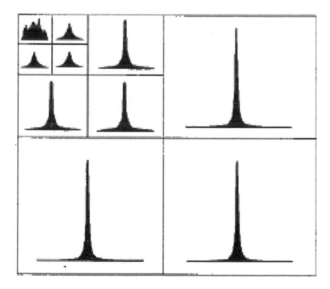

Fig. 5. Histogram of three level wavelet transform of an image (Kilic & Yilmaz, 2003).

In summary, it can be concluded that distribution of the wavelet coefficients in high frequency subbands of natural images can be well approximated by a Gaussian distribution. Therefore, effective use of statistical parameters of the transformed image data (mean values and variances of a Gaussian distribution function) is key in estimation of the transformed data and yielding compression.

4.2 Statistical encoder

In section 4.1 the Gaussian distribution function and its statistical parameters were reviewed. It was shown that every Gaussian distribution function is defined by two parameters: the mean value, which defines the central location of the distribution, and the variance, which defines the width of the distribution. It was also noted that the distribution of the coefficients in each detail subband of the wavelet-transformed data of the natural images is approximately Gaussian has led to the development of a Statistical Encoding (SE) algorithm. The SE algorithm assumes that the coefficients in the 2D input matrix partly follows the Gaussian distribution. Therefore it estimates those parts through a novel hierarchical estimation algorithm, which codes in a lossy manner those parts with their mean values. The SE algorithm applies a threshold value on the variance of the input data to determine if it is possible to estimate them with the mean value of a single Gaussian distribution function or if it needs further dividing into four sub-matrices. This hierarchal algorithm is iterated on the resulting sub-matrices until the distribution of the coefficients in all sub-matrices fulfils the above criteria. Finally, the SE algorithm takes the Gaussian mean values of the resulting sub-matrices as the estimation value for those sub-matrices. The SE algorithm generates a quadtree-like binary map along with the mean values to keep a record of the location of the sub-matrices, which are estimated with their mean values.

A block diagram of the SE algorithm is shown in Figure 6. A two dimensional matrix of size N×N, which for simplification is called U, along with a threshold value, which represents the level of compression, are input to the SE technique. The SE algorithm performs the following process to compress the input matrix U: The SE algorithm first defines two empty vectors called mv (mean value vector) and q (quadtree-like vector). It then calculates the variance (var) and the mean value (m) of the matrix U and compares the resulted variance value with the threshold value. If the variance is less than the threshold value, the matrix is coded by its mean value (m) and one bit binary data equal to 0, which are placed in the mv and q vectors, respectively. If the variance is greater than the threshold, one bit binary data equal to one is placed at the q vector and the size of the matrix is checked. If the size of the matrix is 2×2, the four coefficients of the matrix are scanned and placed in the mv vector and encoding process is finished by sending the mean value vector mv and the quadtree-like vector q. If the size of the matrix is greater than 2×2, the matrix U is divided into four equal non-overlapping blocks. These four blocks are then processed from left to right, as shown in Figure 6. For simplify, only the continuation of the coding process of the first block, U1, is discussed. This process is repeated exactly on the three other blocks. Processing of the first block U1 is described as follows: The variance (var1) and the mean value (m1) of the sub-matrix U1 are first calculated and then the resulting variance value is compared with the input threshold value. If it is less than the threshold value, the calculated mean value (m1) is concatenated to the mean value vector mv and one bit binary data equal to 0 is appended to the quadtree-like vector q. The encoding process of this sub-block is terminated at this stage. Otherwise, the size of the sub-block is checked. If it is 2×2, one bit binary data equal to 1 is appended to the current quadtree-like vector q and the four coefficients of the sub-block are scanned and concatenated to the mv vector and encoding process is ended for this sub-block. If its size is larger than 2×2, one bit binary data equal to 1 is concatenated to the current quadtree-like vector q and the sub-block U1 is then divided into four equal non-overlapping blocks. These four new sub-blocks are named successor sub-blocks and are processed from left to right in the same way that their four ancestor sub-blocks were encoded. The above process is continued until whole successor blocks are encoded. When the encoding process is finished two vectors **mv** and **q** represent the compressed data of the input matrix U.

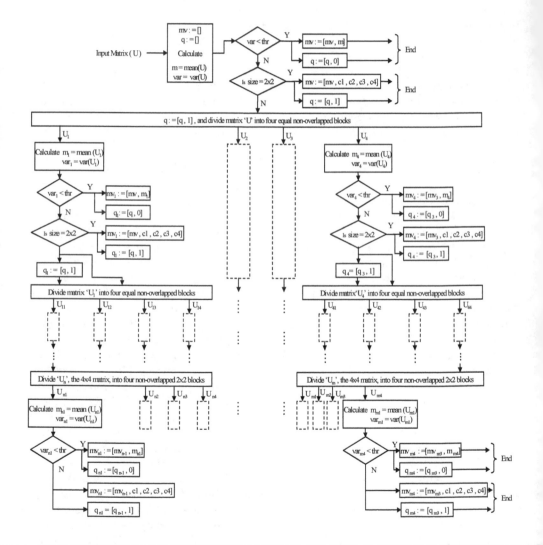

Fig. 6. Block diagram of the Statistical Encoder.

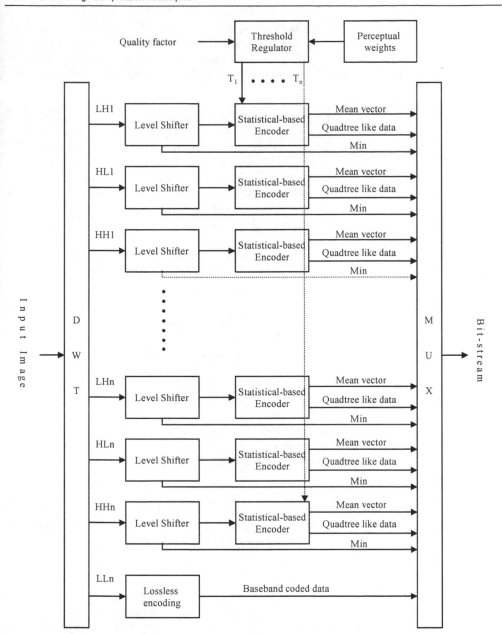

Fig. 7. The multi-resolution and statistical based image encoder.

4.3 Statistical and wavelet based image codec

A block diagram of the Multi-resolution and Statistical Based (MSB) image-coding algorithm is shown in Figure 7. A gray scale image is input to the image encoder. The MSB encoder then applies a 2D lifting based Discrete Wavelet Transform (DWT) to the input image data and decomposes them into a number of subbands. The DWT concentrates most of the image energy into the baseband. Hence, the baseband is losslessly coded using a Differential Pulse Code Modulation (DPCM) algorithm, which will be explained at the end of this section, to preserve visually important information in the baseband. Coefficients in each detail subband are coded using the procedure that is illustrated in Figure 7 as follows: (i) The coefficients in each detail subband are first level shifted to have a minimum value (Min) of zero; (ii) The resulting level shifted coefficients are then coded using the SE algorithm. The SE algorithm takes the level shifted coefficients of a detail subband and a threshold value, which is specifically designed for that subband, and performs the encoding process (The procedures for generating threshold values for different subbands are explained in Section 4.2.1); (iii) The output of each SE encoder is a mean value vector (mv), which carries the mean values, and a quadtree-like vector (q), which carries the quadtree-like data; (iv) Finally the multiplexor combines all the resulting data together and generates the compressed output bitstream.

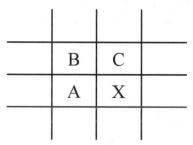

Fig. 8. Three-sample prediction neighbourhoods for DPCM method.

In the DPCM method pixel X with the value of x, is predicted from its three neighbouring pixels, called: A, B and C, with the values of a, b and c respectively, as shown in Figure 8. The prediction value of pixel X, called Px, is calculated using Equation 14:

$$P_x = b + \frac{a-c}{2}$$

(14)

The predicted value of pixel X is then subtracted from the actual value of pixel X to generate an error value, and all the resulting error values are finally losslessly coded.

4.3.1 Threshold generation

In this research work, perceptual weights are employed to regulate the threshold values for different subbands. Hence, the threshold value for each detail subband is generated using a uniform quality factor divided by the perceptual weight of the centre of that subband, where the uniform quality factor can take any positive value. There is a direct relationship between the uniform quality factor and the resulting compression ratios. In Section 3.2 an algorithm for calculating the perceptual weights for detail subbands of a wavelet transformed image data was given. The proposed algorithm is used to calculate the perceptual weights for the centre of each detail subband of an image of size 512×512 and a viewing distance of 40 centimetres, which are shown in Table 6.

DWT Level	Subband	Y-Domain	I-Domain	Q-Domain
	LH	3.0230	1.3251	0.7258
ONE	HL	2.0443	1.0275	0.7681
	HH	0.8713	0.4273	0.4697
	LH	5.4726	2.9355	1.6570
TWO	HL	5.5166	2.3270	1.6560
	HH	2.4531	1.0992	0.8321
	LH	6.1930	4.2479	4.9906
THREE	HL	6.3060	2.9823	4.0070
	HH	4.8143	2.2390	1.6068

Table 6. Perceptual weights for the YIQ colour domain (512×512 image size and a viewing distance of 40 cm).

4.3.2 Results

In order to evaluate the performance of the proposed MSB codec two sets of experiments were performed. In the first sets of experiments the performance of the MSB codec using perceptual weights is compared to that of MSB without using perceptual weights to regulate the threshold values for different subbands, which are presented in Sub-section 4.3.2.1 In the second sets of experiments, the MSB codec using perceptual weights is compared to those of JPEG and JPEG2000 standard image codecs, where the results are illustrated in Sub-section 4.3.2.2.

4.3.2.1 Results for the codec with and without using perceptual weights

The performance of the MSB image codec was investigated on three greyscale test images (with resolution of 8-bits per pixel) and size of 512×512 pixels: 'Lena', 'Elaine', and 'House'. These test images cover all range of spatial frequencies from very low frequency smooth areas, to textures with middle frequencies, and very high frequency sharp edges. In order to evaluate the effect of the perceptual weights on the performance of the proposed codec, 'Lena', 'Elaine', and 'House' test images were compressed using the proposed codec with and without using perceptual weights to regulate the uniform threshold value for different subbands. A three level Daubechies 9/7 wavelet transform was used to decompose the input image into ten subbands for this experiment. The PSNR criterion was used to evaluate the quality of the reconstructed images. The PSNR measurements for the test images at different compression ratios using the MSB codec with and without perceptual weights are given in Figure 9(a) to 9(c) respectively. From these Figures, it is clear that the MSB codec using perceptual weights gives significantly higher performance to that of the MSB codec without perceptual weights. However, it is well known that the PSNR is an unreliable metric for measuring the visual quality of the decompressed images (Kaia et al., 2005). Hence, to illustrate the true visual quality obtained using the MSB codec with and without perceptual weights, the reconstructed 'Lena', 'Elaine', and 'House' images at compression ratio of 16 using the proposed codec are shown in Figure 10(I) to 10(III), respectively. From theses figures, it can be seen that the reconstructed images, when perceptual weights are used in the encoding process, have significantly higher visual quality with less blurred edges and better surface details. From Figure 10(I) and 10(II), which show decoded 'Lena' and 'Elaine' test images, it is obvious that the images using the MSB codec using perceptual weights have a noticeably higher quality to those decoded using the MSB codec without employing perceptual weights. It can also be seen that the decoded test images using the MSB codec with perceptual weights have clearer facial details with less blurring in the faces. From Figure 10(III) it is clear that the reconstructed 'House' test image using MSB with HVS have significantly higher visual quality with lower blurred edges and clearer surface details.

4.3.2.2 Results of the MSB, JPEG and JPEG2000 codecs

In this section, the performance of the MSB codec with perceptual weights is compared to JPEG and JPEG2000 (JPEG2000, 2005) standard image coding techniques. The MSB, JPEG and JPEG2000 were used to compress "Lena', 'Elaine', and 'House' test images at different compression ratios. The PSNR measurements for the encoded images using the MSB, JPEG, and JPEG2000 image codecs at different compression ratios are shown in Figures 11(a) to 11(c), respectively. From these figures it can be seen that the MSB codec gives superior performance to JPEG and JPEG2000 at low compression ratios. From Figure 11(a) and 11(b), it can be observed that the proposed codec offers higher PSNR in coding 'Lena' and 'Elaine' test images to those of JPEG and JPEG2000 at compression ratios lower than 5. From Figure 11(c), it is clear that the MSB codec outperforms JPEG and JPEG2000 in coding 'House' test images at compression ratios of up to 4. However, it is well known that the PSNR often does not reflect the visual quality of the decoded images, thus a perceptual quality evaluation seems to be necessary. To demonstrate the visual quality achieved using the MSB, JPEG and JPEG2000 coding techniques at different compression ratios, the decoded 'Lena' and 'Elaine' test images at compression ratios 5 and 40 using these techniques are shown in Figures 12 and 13, respectively.

From Figures 12(a), it can be seen that the visual quality of the decoded Lena test image at a compression ratio of 5 using MSB codec is high. It is also clear that the quality of the decoded Lena test image using MSB codec is slightly higher than that of JPEG and almost the same as that of JPEG2000. The Elaine test image contains significant high frequency details and is more difficult to code. From Figure 12(b), which illustrates the decoded Elaine test images at compression ratio of 5, the high visual quality of all the decoded images is obvious. From Figures 13(a), which illustrates the decoded Lena test images at a compression ratio of 40, the severe blocking artefact of the decoded image using JPEG is quite obvious, where the MSB decoded image contains some blurring around the mouth and ringing artefacts around edges in the image. In terms of overall visual quality, the MSB decoded Lena test image has superior visual quality to that of JPEG. It is also clear that the quality of the MSB decoded image is slightly inferior to that of JPEG2000. From Figures 13(b), which illustrates the decoded Elaine test images at a compression ratio of 40, it is obvious that: a) the decoded JPEG image exhibits severe blocking artefacts; b) the MSB decoded image has higher visual quality but suffers from blurring in the background of the image and ringing artefacts in its sharp edges and c) the JPEG2000 decoded image has high visual quality but slight blurring and ringing artefacts can be seen in some regions of the background and sharp edges of the image. It is clear that the JPEG2000 decoded images have slightly higher visual quality than MSB decoded images.

The results presented here demonstrate that the MSB codec outperforms JPEG and JPEG2000 image codecs, subjectively and objectively, at low compression ratios (up to compression ratio of 5). The results also show that at middle-range compression ratios JPEG decoded images somewhat suffer from blocking artefacts, while the visual quality of the MSB decoded images is significantly higher.

The results at high compression ratios (around 40) indicate that a) the JPEG decoded images severely suffer from blocking artefacts, so much so that there is no point in using JPEG to code images at high compression ratios; b) the MSB decoded images have significantly higher visual quality than that of JPEG, while they slightly suffer from patchy blur in regions with soft texture and ringing noise at sharp edges; c) decoded MSB images have significantly lower PSNR in comparison to that of JPEG2000 but their visual quality is slightly inferior to that of JPEG2000.

5. Conclusion

In this Chapter first a novel statistical encoding algorithm was presented. The proposed SE algorithm assumes that the distribution of the coefficients in the input matrix is partly Gaussian and uses a hierarchal encoding algorithm to estimate the coefficients in the input matrix with the Gaussian mean values of multiple distributions; then a multi-resolution and statistical based image-coding scheme was developed. It applies a 2D wavelet transform on the input image data to decompose it into its frequency subbands. The baseband is losslessly coded to preserve the visually important image data. The coefficients in each detail subband were first dc level shifted to have a minimum value of zero and then coded using the SE algorithm. The SE algorithm takes the dc level shifted coefficients of a detail subband and a threshold value, which is generated for that subband. The encoding process is then performed. Perceptual weights were calculated for the centre of each detail subband and used to regulate the threshold value for that subband.

Experimental results showed that the proposed coding scheme provides significantly higher subjective and objective quality when perceptual weights are used to regulate the threshold values. The results also indicated that the proposed codec outperforms JPEG and JPEG2000 coding schemes subjectively and objectively at low compression ratios. Results showed that the proposed coding scheme outperforms JPEG subjectively at higher compression ratios. It offers comparable visual quality to that of JPEG2000 at high compression ratios.

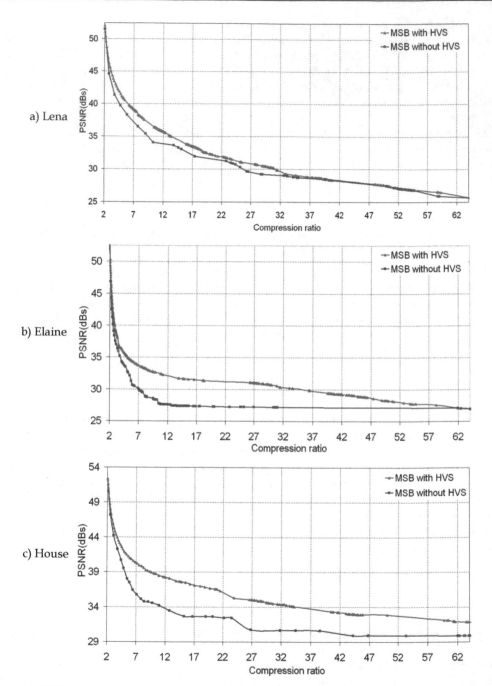

Fig. 9. PSNR measurements for a) 'Lena', b) 'Elaine' and c) 'House' test images at different compression ratios using MSB codec with and without employing perceptual weights.

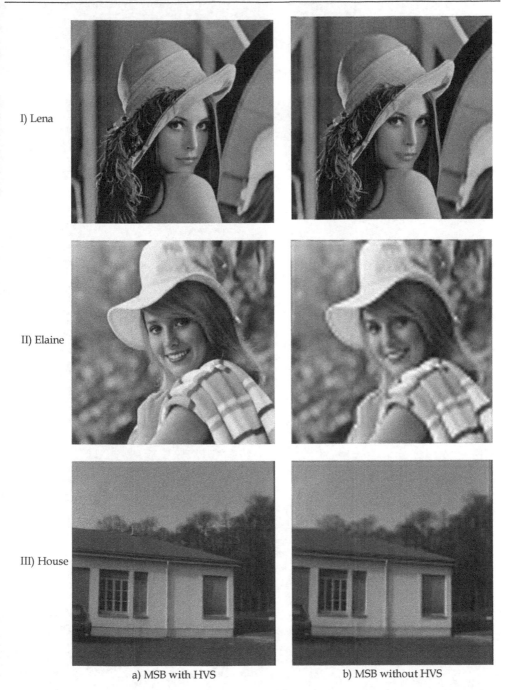

I) Lena

II) Elaine

III) House

a) MSB with HVS   b) MSB without HVS

Fig. 10. Reconstructed I) 'Lena', II) 'Elaine' and II) 'House' test images at compression ratio of 16 using the MSB codec a) with HVS and b) without HVS.

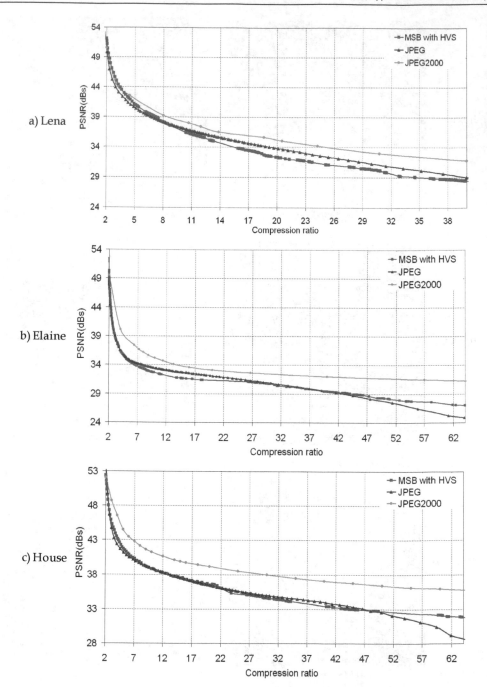

Fig. 11. PSNR measurements for a) 'Lena', b) 'Elaine' and c) 'House' test images at different compression ratios using MSB, JPEG and JPEG2000 codecs.

a) Lena   b) Elaine

Fig. 12. Reconstructed a) 'Lena' and b) 'Elaine' test images at compression ratio of 5 using I) MSB codec, II) JPEG and III) JPEG2000.

I) MSB

II) JPEG

III) JPEG 2000

a) Lena   b) Elaine

Fig. 13. Reconstructed a) 'Lena' and b) 'Elaine' test images at compression ratio of 40 using I) MSB codec, II) JPEG and III) JPEG2000.

6. References

(Aili et al., 2006) W. Aili, Z. Ye and G. Yanfeng, "SAR Image Compression Using HVS Model", International Conference on Radar 2006, pp. 1-4, Oct. 2006.

(Altunbasak & Kamaci, 2004) Y. Altunbasak and N. Kamaci, "An Analysis Of The DCT Coefficient Distribution with The H.264 Video Coder", *ICASSP2004*, pp.178-180, 2004.

(Antonini et al., 1992) M. Antonini, M Barlaud, P. Mathieu and I. Daubechies, "Image coding using the wavelet transform," *IEEE Transaction on Image Processing*, Vol. 1, No. 2, pp. 205-220, April 1992.

(Bradley, 1999) A. P. Bradley, A wavelet visible difference predictor, *IEEE Transaction on Image Processing*, vol.8, no. 5, 1999.

(Campbell & Robson, 1968) F.W. Campbell and J. G. Robson, "Application of Fourier Analysis to the Visibility of Gratings", *Journal of Physiologic*, vol. 197, pp. 551–566, 1968.

(Eude et al., 1994) T. Eude, H. Cherifi and R. Grisel, "Statistical distribution of DCT coefficients and their application to an adaptive compression algorithm", *IEEE International Conference TENCON1994*, pp. 427-430, 1994.

(Ghanbari, 1999) M. Ghanbari, "Video coding an introduction to standard codecs," *Published by: The Institution of Electrical Engineering, London, UK*, 1999.

(Chang & Chen, 1993) R. F. Chang and W. T. Chen, "Image Coding Using Variable-Rate Side –Match Finite-State Vector Quantization," *IEEE Transaction on Image Processing*, vol. 2, no. 1, January 1993.

(HSontsch & Karam, 2000) I. Höntsch and L. Karam, "Locally adaptive perceptual image coding," *IEEE Transaction on Image Processing*, vol. 9, no. 9, pp. 1472–1483, September 2000.

(JPEG, 1994) Information Technology-JPEG-Digital Compression and Coding of Continuous-Cone Still Image-Part 1: Requirement and Coding, 1994. ISO/IEC 10918-1 and ITU-T Recommendation T.81.

(JPEG2000, 2000) JPEG 2000 Part I Final Committee Draft Version 1.0, ISO/IEC JTC1/SC29/WG1 N1646R, Mar. 2000.

(Kilic & Yilmaz, 2003) I. Kilic and R. Yilmaz, "A Video Compression Technique Using Zerotree Wavelet and Hierarchical Finite State Vector Quantization" Proceedings of the 3rd International Symposium on Image and Signal Processing and Analysis2003 (ISPA2003), pp. 311-316, 2003.

(Lu et al., 2000)] Z. M. Lu, J. S. Pan and S. H. Sun, "Image Coding Based On Classified Side-Match Vector Quantization," *IEICE Transaction Information System*, E83-D, no. 12, December 2000.

(Lu et al., 2002) Z. M. Lu, B. Yang and S. H. Sun, "Image Compression Algorithms Based On Side-Match Vector Quantization With Gradient-Based Classifiers," *IEICE Transaction Information System*, E85-D, no. 9, September 2002.

(Nadenau et al., 2003) M. J. Nadenau, J. Reichel and M. Kunt, "Wavelet-Based Colour Image Compression: Exploiting the Contrast Sensitivity Function", *IEEE Transaction on Image Processing*, vol. 12, no. 1, pp. 58-70, January 2003.

(Ostermann et al., 2004) J. Ostermann, J. Bormans, P. List, D. Marpe, M. Narroschke, F. Pereira, T. Stockhammer, and T. Wedi, "Video coding with H.264/AVC: Tools, Performance, and Complexity," *IEEE Circuits and System Magazine*, Vol. 4, No. 1, pp. 7-28, First Quarter 2004.

(Peng & Kieffer, 2004) K. Peng and J. C. Kieffer, "Embedded Image Compression Based on Wavelet Pixel Classification and Sorting," *IEEE Transaction on Image Processing*, vol. 13, no. 8, pp. 1011-1017, August 2004.

(Said & Pearlman, 1996) A. Said and W. Pearman, "A new, Fast, and Efficient Image Codec Based on Set Partitioning in Hieratical Trees", *IEEE Transaction on Circuits and Systems for Video Technology*, Vol. 6, no. 3, pp. 243-250, June1996.

(Saryazdi & Jafari, 2002) S. Saryazdi and M. Jafari, "A High Performance Image Coding Using Uniform Morphological Sampling, Residues Classifying, and Vector Quantization," *EurAsia-ICT 2002*, Shiraz, Iran, October 2002.

(Scargall & Dlay, 2000) L. D. Scargall and S. S. Dlay, "New methodology for adaptive vector quantization", *IEE Proceedings on Vision, Image and Signal Processing*, Vol. 147, No. 6, December 2000.

(Shapiro, 1993) J.M Shapiro, "Embedded image coding using zerotrees of wavelet coefficients", *IEEE Transactions on Acoustics, Speech, and Signal Processing*, vol. 41, Issue 12, pp. 3445 - 3462, 1993.

(Sheikh Akbari & Soraghan, 2003) A. Sheikh Akbari and J.J. Soraghan, "Adaptive Joint Subband Vector Quantization Codec For Handheld Videophone Applications", *International Journal of IEE Electronic Letters*, VOL. 39. NO.14, pp. 1044 – 1046, 2003.

(Skodras et al, 2001) A. Skodras, Ch. Christopoulos, and T. Ebrahimi, "The JPEG 2000 Still Image Compression Standard," *IEEE Signal Processing Magazine*, vol. 18, no.5, September 2001.

(Tan et al, 2004) D. M. Tan, H. R. Wu, and Z. Yu, "Perceptual Coding of Digital Monochrome Images," *IEEE Signal Processing Letters*, Vol. 11, no. 2, pp. 239-242, February 2004.

(Thornton et al., 2002) L. Thornton, J. Soraghan, R. Kutil, M. Chakraborty, "Unequally protected SPIHT video codec for low bit rate transmission over highly error-prone mobile channels," *Elsevier Science: Signal Processing: Image Communication*, vol. 17, pp. 327–335, 2002.

(Valade & Nicolas, 2004) C. Valade, J. M. Nicolas, "Homomorphic wavelet transform and new subband statistics models for SAR image compression" IEEE International Proceedings on Geoscience and Remote Sensing Symposium, IGARSS 2004, Vol.1, 2004.

(Van Dyck & Rajala, 1994) R.E. Van Dyck, and S. A. Rajala, "Subband/VQ Coding of Colour Images with Perceptually Optimal Bit Allocation", *IEEE Trans. on Circuits and Systems for Video Technology*, vol. 4, no. 1, February 1994.

(Voukelatos & Soraghan, 1997) S. P. Voukelatos and J. J. Soraghan, "Very Low Bit Rate Color Video Coding Using Adaptive Subband Vector Quantization with Dynamic Bit Allocation", *IEEE Transaction on Circuits and Systems for Video Technology*, vol. 7, no. 2, pp. 424-428, April 1997.

(Voukelatos & Soraghan, 1998) S. P. Voukelatos and J. J. Soraghan, "A multiresolution adaptive VQ based still image codec with application to progressive image transmission", *EURASIP Signal Processing: Image Communication*, vol. 13, no. 2, pp. 135-143, 1998.

(Wang et al., 2001) J. Wang, W. Zhang and S. YU, "Wavelet coding method using small block DCT," *Electronic Letters*, Vol. 37, No. 10, pp.627-629, May 2001.

(Yovanof & Liu, 1996) Yovanof and S. Liu, "Statistical analysis of the DCT coefficients and their quantization error", Thirtieth Asilomar Conference on Signals, Systems and Computers, pp. 601-605, 1996.

3

Information Extraction and Despeckling of SAR Images with Second Generation of Wavelet Transform

Matej Kseneman[1] and Dušan Gleich[2]
[1]Margento R&D d.o.o.,
[2]University of Maribor, Faculty of Electrical Engineering and Computer Science,
Slovenia

1. Introduction

Synthetic Aperture Radar (SAR) technology is mainly used to obtain high-resolution images of ground areas in resolutions even less than meter. SAR is even capable of imaging a wide area of terrain and from two and more images it is possible to reconstruct a 3D digital elevation model of ground terrain. Good thing about SAR is an all whether operation and possibility to capture images under various inclination angles. Because digital images are usually corrupted by noise that arises from an imaging device, there is always a need for a good filtering algorithm to remove all disturbances, thus enabling more information extraction. The SAR images are corrupted by a noise called speckle, which makes the interpretation of SAR images very difficult. The goal of removing speckles from the SAR image is to represent a noise-free image and preserve all important features of the SAR image, as for example edges, textures, region borders, etc.

Many different techniques for SAR image despeckling have been proposed over the past few years. Speckle is a noise-like characteristic of SAR images and it is a multiplicative nature, if the intensity or amplitude image is observed. The despeckling can be performed in the image or in the frequency domain. The well-known despeckling filters are Lee (Lee, 1980), Kuan (Kuan et al., 1985), and Frost (Frost et al., 1982). Lee and Kuan filters can be considered as an adaptive mean filters, meanwhile the Frost filter can be considered as a mean adaptive weighted filter. The Bayesian filters are based on the Bayesian theorem, which defines a posterior probability by using a prior, likelihood and evidence probability density functions (pdf). The solution for noise-free image is found by a maximum a posteriori (MAP) estimate. The MAP estimate of a noise free image was proposed in (Walessa & Datcu, 2000), where the noise free image was approximated by a Gauss-Markov random field prior and the noise was modeled with Gamma pdf. Model based despeckling and information extraction is one of the promising techniques of SAR image denoising and scene interpretation. The wavelet based despeckling algorithms have been proposed in (Dai et al., 2004), (Argenti & Alparone, 2002), and (Foucher et al., 2001). The second generation wavelets Chirplet (Cui & Wong, 2006), Contourlet (Chuna et al., 2006), Bandelet (Le Pennec & Mallat, Apr 2005) have appeared over the past few years.

First transform we used is so called Bandelet transform (Le Pennec & Mallat, Dec 2005), which further divides wavelet subbands into smaller subbands using a rate distortion optimization that enables removing redundancy in wavelet transformation. Bandelets (Le Pennec & Mallat, Dec 2005) contain anisotropic wavelets which combine redundancy in the geometric flow of an image corresponding to local directions of its grey levels. With this geometric flow wavelet warping represents a vector field with indication of regularity along edges. Bandelet decomposition is constructed in much the same way as wavelet with use of dyadic squares containing information about bandelet coefficients (parameterized geometric flow) and segmentation (Le Pennec & Mallat, Apr 2005). These squares summarize geometry by local clustering of similar directional vectors. A Bandelet transform can be viewed as an adaptive wavelet basis transform, which is warped according to local direction.

Bandelet transform is therefore capable to separate two different surface areas with different curvatures, which are then decomposed into optimal estimations of regularity direction (Le Pennec & Mallat, Apr 2005). The geometry itself is obtained with regularity flow estimation. Fig. 1 shows an example of directions acquired with bandelet transform. The computational complexity of this transform is much higher as in the case of the classical dyadic decomposition.

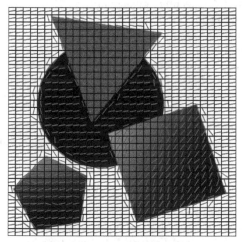

Fig. 1. Directions obtained by bandelet transform

The contourlet transform (Do & Vetterli, 2005) is organized a little bit different, because this transform is directly constructed in a discrete space. Thus, contourlet does not need to be transformed from continuous time-space domain. In order to capture as much as possible directional information a 2D directional filter bank is used in contourlet transform (Do & Vetterli, 2005). Directional filter bank is represented with k-binary tree which decomposes original image into $2k$ bands. These directional filter banks have a flaw mainly because they are designed to capture a direction, which is mainly done in high frequency spectrum of the input image, therefore low frequencies are obstructed. Low frequencies can easily penetrate into several different directional subbands, thus corrupting the transformation subbands. To solve this problem a multiscale decomposition is created with directional decomposition with the help of Laplacian pyramid as a low frequency filter. Laplacian pyramid throughput is a band-pass image which is then led to directional filter bank where a directionality of an image is captured. This scheme can be further applied on a coarser image and thus an iterative scheme can be achieved. It can be concluded that applying iterative contourlet scheme derives to directional subbands in a presence of multiple different scales.

For the despeckling of TerraSAR-X (Wikipedia, 2011) images we used a model based approach, which is supported by first order Bayesian inference. After applying transforms to images a general Gaussian distribution appears in wavelet domain. In this wavelet domain we get subbands different in scales and frequency. The subbands in the wavelet domain have Gaussian distribution and therefore the general Gaussian model is used for a prior density function (pdf). The likelihood pdf is modeled using Gaussian pdf in both, bandelet and contourlet transforms. The despeckling using contourlet (Li et al., 2006) and bandelet (Sveinsson et al., 2008) transforms showed superior despeckling results for SAR images comparing with the wavelet based methods. The model based despeckling mainly depends on chosen models. The image and noise models in the wavelet domain are well defined with presented results in (Argenti & Alparone, 2002), (Gleich & Datcu, 2006) and usually noise-free image is computed using maximum a posteriori estimate.

The despeckling methods were tested using synthetic and real TerraSAR-X data, which were captured in the high resolution spotlight mode. The experimental results showed that the best despeckling method for synthetic images is bandelet transform, because contourlet transform produces artifacts in the homogenous areas. The ratio images between original and despeckled images were examined in order to show estimation of speckle noise, edge and texture preservation using bandelet and contourlet transform. The contourlet transform produces artifacts in form of lines in both homogenous areas and edges.

2. Second generation wavelets

In this section a comparison between bandelets and contourlets is presented. Bandelets and contourlets are presented in great detail, including subbands creation and filter decomposition. These two denoising schemes are a foundation of later proposed model, which builds a denoising scheme on top of these two schemes yielding better denoising results.

2.1 Bandelets

Bandelets (Le Pennec & Mallat, Apr 2005), (Le Pennec & Mallat, Dec 2005) belong to a second generation of wavelet transforms and are composed of anisotropic wavelets, which are in fact a combination of geometric flow of an image corresponding to local directions of its gray levels. This geometric flow represents a regularity of a vector field along edges contained in the image. Typical example of this geometric flow can be seen on Fig. 1, where it can be observed that all directions are aligned to object's edges at the boundary of two different areas.

Edges inside an image are often hard to determine. First generation of bandelet transform uses the vector field (Le Pennec & Mallat, 2001), which determines image regularities and irregularities. Therefore bandelet coefficients represent geometric flow defined by polynomial function. This geometric flow consists of directions of variations in image grey levels, where linear geometric flow is preferred. Bandelet transform image is divided into regions with corresponding vector fields, which describes directions of regularity inside a predefined neighborhood.

If the image intensity is uniformly regular in the neighborhood of a point, then this direction is not uniquely defined, and some form of global regularity is therefore imposed on the flow to specify it uniquely.

In literature it has been proven that the first generation of bandelets has minimum distortion for images whose edges correspond to geometric regularity. However, the first generation of bandelets is composed in continuous space, thus not being able to represent a multi-resolution of the geometric regularity. Thus, the second generation of bandelets (Le Pennec & Mallat, Apr 2005) was introduced, which is an orthogonal multiscale transform constructed directly in discrete domain. The bandelet transform first creates a composition of smaller images representing subbands, and then uses fast subband-filtering algorithms. For applications including speckle-noise removal, the geometric flow is optimized in a way that bandelet transform produces minimum distortion in reconstructed images. The decomposition on a bandelet basis is computed using a wavelet filter bank followed by adaptive geometric orthogonal filters, which require $O((\log_2 N)^3)$ operations.

The key parameters in bandelet transform are: the estimation of basis shapes, the partition of images, and the optimization of geometric flows (Yang et al., 2007). To represent image with as little as possible information, the complex edges must be divided into simpler smaller shapes so that linear geometric flows can represent them sufficiently. The image is commonly divided into smaller square regions that are being divided until there is only one contour inside a square region. It must be noted that the geometric regularity should be discrete, so dyadic decomposition by successive subdivisions of square regions into four smaller sub-squares of twice smaller width can be made. There is a defined maximum and minimum block size (Le Pennec & Mallat, Apr 2005), where the first division produces blocks of maximum size, while later iterations divide those blocks up until minimum size is reached. This partition result can be viewed as a quadtree, where each block is represented by its corresponding leaf in a tree. At each scale the resulting geometry is multiscale and calculated by a procedure that minimizes the Lagrangian cost function.

Implementation

The bandelet transform first computes the 2D wavelet transform of the input original image (Peyré & Mallat, Apr 2005). This transform is based on orthogonal or biorthogonal filter banks and results in four smaller images (children) containing low- and high-frequency components. By selecting a dyadic square and recursively splitting input wavelet image four new sub-squares are created. Further on geometric flow parameterization is performed in each of these sub-squares in every possible direction. Let us assume that each of these squares has a width of k pixels then the number of potential directions d is a little less than $2k^2$. The sampling location is then projected along potential direction and afterwards sorting the resulting 1D points is performed from left to right direction. These points define 1D discrete signal f_d (Le Pennec & Mallat, Apr 2005) which is later on transformed with 1D discrete wavelet transform. For a given user defined threshold T, the bandelet transform has to find the best available direction, which in fact produces the less approximation error. Best geometry is obtained by choosing best direction d that minimizes the Lagrangian

$$\mathscr{L}(f_d, R) = \|f_d - f_{dR}\|^2 + \lambda T^2 (R_G + R_B) \tag{1}$$

where f_{dR} is the recovered signal from quantized coefficients acquired by inverse 1D wavelet transform, R_G is the number of bits needed to code geometric parameter d, R_B is the number of bits needed to code the quantized coefficients and $\lambda = 3/28$ (Le Pennec & Mallat, Apr 2005).

When there are gathered all approximations over each individual dyadic square, the quadtree can be constructed. The algorithm starts with the smallest squares that represent a leaf in quadtree and initialize the cumulative Lagrangian of the sub-tree. Within these dyadic squares, a best bandelet approximation is obtained by minimizing a Lagrangian cost function (Le Pennec & Mallat, Dec 2005). Fig. 2 shows an example of denoising obtained with the bandelet transform including dyadic squares that indicate a progress of dyadic levels. This image is represented by indexing a dyadic level used in bandelet transform, where white indicates the first level and black the last level achieved.

a)   b)   c)

Fig. 2. An example of image denoising using a bandelet transforms. a) Original image, b) Denoised image using a bandelet transforms, and c) Dyadic squares tree

2.2 Contourlets

Contourlet transform (Do & Vetterli, 2005) is also classified as a second generation wavelet transform for which a Fourier transform is not needed anymore. Main advantage of second generation wavelet transform over the first generation is a true discrete 2D transformation, which is able to capture geometry of an image, but the first generation wavelet transform does not perform very well on edge regions. This transformation therefore results in adaptive multi-resolution and directional image expansion using contour segments.

Best performance of wavelets is achieved in 1D case which is for example only one row of a 2D picture. Because pictures are not simply stacks of rows, discontinuities evolve along smooth regions. 2D wavelet transform thus captures edge points, but on the other hand the throughput on smooth regions is not quite as good anymore. Moreover wavelet transform can only capture a fraction of image directionality, which is clearly seen in Fig. 3 where wavelet transform needs a lot more subdivisions and information then a contourlet transform.

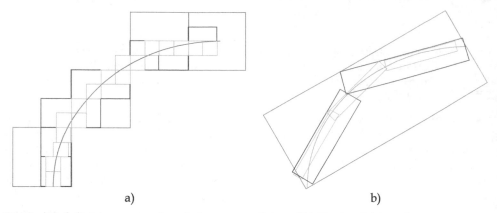

a)   b)

Fig. 3. a) Subdivision comparison between wavelet, and b) Contourlet transform

In order to capture as much as possible directional information a new type of filter bank has to be constructed. Thus a 2D directional filter bank (Bamberger & Smith, 1992) is used in Contourlet transform. Directional filter bank is represented by k-binary tree, which decomposes original image into 2^k subbands as represented in Fig. 4. The algorithm based on contourlet transform uses a simpler version of directional filter bank, where the first part is constructed from two-channel quincunx filter bank (Vetterli, 1984), while the second part is sampling and reordering operator. With this composition a frequency partition is achieved and also a perfect reconstruction is obtained. As shown in Fig. 6 one can obtain different 2D spectrum decompositions with appropriate combinations of aforementioned building blocks. Thus a k-level binary tree directional filter bank can be viewed as 2^k parallel channel filter bank with equivalent filters and its sampling matrices as shown in Fig. 5 (Do & Vetterli, 2005). In Fig. 5 D denotes an equivalent directional filter.

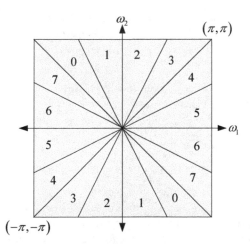

Fig. 4. Frequency partitioning where $k = 3$ and there are $2^3 = 8$ real wedge-shaped frequency bands. Subbands 0-3 correspond to the mostly horizontal directions, while subbands 4-7 correspond to mostly vertical directions

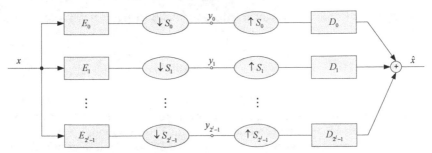

Fig. 5. The multichannel view of a k-level tree-structured directional filter bank

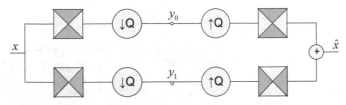

Fig. 6. 2D partition of spectrum using quincunx filter banks with fan filters. Darker shades represent the ideal frequency supports of each filter. Denotion Q represents a quincunx sampling matrix

These directional filter banks have a flaw mainly because they are designed to capture directions, which is mainly done in high frequency spectrum of the input image and thus low frequencies are obstructed. As Fig. 5 shows frequency partition a low frequencies can easily penetrate into several different directional subbands and therefore corrupt the transformation subbands. It is therefore wise to combine multiscale decomposition with directional decomposition with the help of Laplacian pyramid as a low frequency filter. Laplacian pyramid throughput is a bandpass image which is then led to directional filter bank where a directionality of the image is captured. This scheme can be further applied on a coarser image and thus an iterative scheme can be achieved. It can be concluded that applying iterative contourlet scheme derives to directional subbands in presence of multiple different scales, which is depicted in Fig. 7.

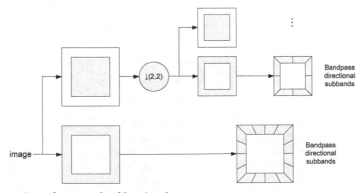

Fig. 7. Construction of contourlet filter bank

3. Bayesian inference incorporated into second generation wavelet transform

The first level of Bayesian inference is given by

$$p(x|y,\theta) = \frac{p(y|x,\theta)p(x|\theta)}{p(y|\theta)} \tag{2}$$

where y represents a noisy image, x represents a noise-reduced image, the θ's are the model parameters, $p(y|x, \theta)$ denotes the conditional pdf called **likelihood**, $p(x|\theta)$ denotes **prior** pdf, and $p(y|\theta)$ represents **evidence** pdf. In Eq. (2), the evidence pdf does not play a role in the maximization over x, and therefore, the MAP estimator can be written by

$$\hat{x}(y) = \arg\max_x p(y|x,\theta)p(x|\theta) \tag{3}$$

where the likelihood and prior pdfs should be defined. The MAP estimator is an optimal estimator and minimizes the given cost function. The speckle noise in SAR images is modeled as multiplicative noise, i.e. $y = x \cdot z$, where z represents pure speckle noise. A multiplicative speckle noise can also be modeled using an additive signal-dependent model, as proposed in (Argenti & Alparone, 2002) $y = x \cdot z = x + x(z - 1) = x + n$, where n is a non-stationary signal-dependent additive noise equal to $x(z - 1)$.

Models describing texture parameters are widely used in SAR image despeckling (Walessa & Datcu, 2000). Let us model the image as generalized Gauss-Markov random fields (GGMRF) given by

$$p(x_s|\theta) = \frac{v\eta(v,\sigma_x)}{2\Gamma(1/v)}\exp\left(-\left[\eta(v,\sigma_x)\left|x_s - \sum_{r\in\zeta_s}\theta_r(x_{s+r}+x_{s-r})\right|\right]^v\right) \tag{4}$$

Neighborhood		
Cliques		
Parameters	θ_1	θ_2

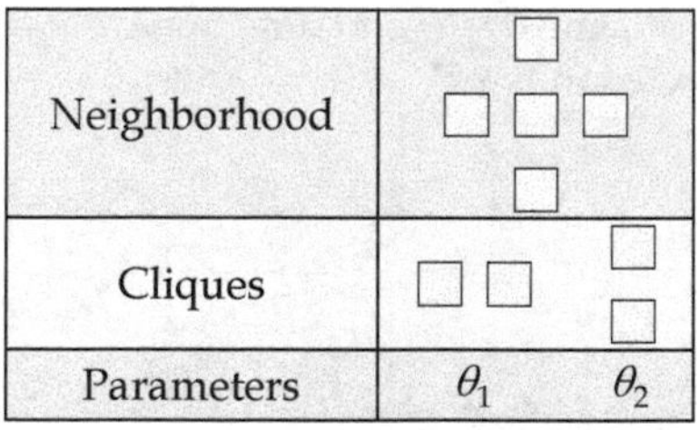

Fig. 8. First order cliques

Let σ_x and θ_s define the GGMRF with a neighborhood set ζ_s. The MRF model characterizes the spatial statistical dependence of 2D data by a symmetric set called neighboring set. The expression $\sum\theta_r(x_{s+r} + x_{s-r})$ in Eq. (4) represents the sum of all the distinct cliques of neighboring pixel at a specific subband level. A clique is defined as a subset of sites neighboring the observed pixel, where every pair of sites is neighbors of each other. In this double site, cliques are used. A sum is performed over horizontal and vertical neighboring pixels, for the first model order of the MRF. The neighbor set for a first model order is defined as $\zeta = \{(0, 1), (0, -1), (1, 0), (-1, 0)\}$, and can be seen in Fig. 8. Moreover, the neighbor set for a second model order is defined as $\zeta = \{(0, 1), (0, -1), (1, 0), (-1, 0), (1, 1), (-1, -1),$ $(1, -1), (-1, 1)\}$. The MRF model is defined for the symmetric neighbor set; therefore, if $r \in \zeta_s$, then $-r \notin \zeta_s$, and ζ is defined as $\zeta = (r: r \in \zeta_s) \cup (-r: r \in \zeta_s)$. The parameter v in (4) represents the shape parameter of the GGMRF, $\Gamma(\cdot)$ represents the Gamma function, and η is given by

$$\eta(v, \sigma_x) = \sigma_x^{-1} \sqrt{\frac{\Gamma(3/v)}{\Gamma(1/v)}} \tag{5}$$

A likelihood pdf is given by a Gaussian distribution

$$p(y_s|x_s) = \frac{1}{\sqrt{2\pi\sigma_n^2}} \exp\left(-\frac{(y_s - x_s)^2}{2\sigma_n^2}\right) \tag{6}$$

where σ_n^2 is a noise variance.

The noise variance σ_n^2 can be estimated using the results presented in (Argenti & Alparone, 2002), and is given by

$$\sigma_n^2 = \psi_l \mu_x^2 C_F^2 \left(1 + C_x^2\right) \tag{7}$$

where $\mu_x = E[x]$, and $E[x]$ denotes a mathematical expectation. C_x^2 is given by

$$C_x^2 = \frac{C_{Wy}^2 - \psi_l C_F^2}{\psi_l \left(1 + C_F^2\right)} \tag{8}$$

The normalized standard deviation of the noisy wavelet coefficient is given by $C_{Wy} = \sigma_{Wy}/\mu_y$, where σ_{Wy} is a standard deviation calculated within the wavelet domain, and μ_y is the mean value calculated in the spatial domain. C_F denotes the normalized standard deviation of the speckle noise. The parameter ψ_l is defined as a product of the coefficients from high-pass (g_k) and low-pass (h_k) filter used at the l-th level of the wavelet decomposition. If the wavelet coefficients of a diagonal detail are of interest, then the parameter ψ_l is given by

$$\psi_l = \left(\sum_k h_k^2\right)^2 \left(\sum_k g_k^2\right)^{2(l-1)} \tag{9}$$

Moreover, if the wavelet coefficients in the horizontal and vertical details are of interest, then the parameter ψ_l is given by

$$\psi_l = \left(\sum_k h_k^2\right)\left(\sum_k g_k^2\right)^{2l-1} \tag{10}$$

Since the random variable z of the speckle noise is normalized (i.e. $E[z] = 1$), the parameter C_F for intensity images is given by

$$C_F = 1/\sqrt{L} \tag{11}$$

while for the amplitude images the parameter C_F is given by

$$C_F = \sqrt{(4/\pi - 1)/L} \tag{12}$$

The parameter L represents the number of looks of the original SAR image. However, its value is unknown, thus an approximation has to be done, which is $L = \mu^2/\sigma^2$. The noise variance is then given by

$$\sigma_n^2 = \frac{C_F^2\left(\psi_l \mu_y^2 + \sigma_{Wy}^2\right)}{1 + C_F^2} \tag{13}$$

where $\mu_y = E[y]$. Noise-reduced variance can be computed using the results presented in the paper (Argenti & Alparone, 2002). Thus, noise-reduced the variance is given by

$$\sigma_x^2 = \psi_l \mu_x^2 C_x^2 \tag{14}$$

Where μ_x^2 is the mean value calculated within the wavelet domain over a predefined window size.

The MAP estimate using the GGMRF primarily defined in (4) and the likelihood defined in (6) is given by

$$-v\eta(v,\sigma_x)\left[\eta(v,\sigma_x)\left|x_s - \sum_{r\in\zeta_s}\theta_r\left(x_{s+r} + x_{s-r}\right)\right|\right]^{v-1} + \frac{y_s - x_s}{\sigma_n^2} = 0 \tag{15}$$

The evidence maximization algorithm is used in order to find the best model's parameters (v, θ). The analytical solution for the integral over the posterior $p(y \mid x)p(x \mid \theta)$ does not exists; therefore, the evidence is approximated. The multidimensional pdf is approximated by the multivariate Gaussian distribution with Hessian matrix H centered on the maximum of the a posteriori distribution (Walessa & Datcu, 2000), (Sivia, 1996). The integral over a posterior pdf consists of mutually independent random variables; therefore, a conditional pdf can be rewritten as a product of their components

$$p(y|\theta) = \int p(y|x)p(x|\theta)dx$$

$$p(y|\theta) \approx \int \prod_{i=1}^{N} p(y_i|\hat{x}_i)p(\hat{x}_i|\theta)\exp\left(-\frac{1}{2}\Delta x^T H \Delta x\right)dx \tag{16}$$

$$p(y|\theta) \approx \frac{(2\pi)^{N/2}}{\sqrt{|H|}}\prod_{i=1}^{N} p(y_i|x_i)p(x_i|\theta)$$

where $\Delta x = x - \hat{x}$ and Hessian matrix H is a square matrix of the second-order partial derivatives of a univariate function

$$H = -\nabla\nabla\sum_{i=1}^{N}\log\left(p(y_i|x_i)p(x_i|\theta)\right) \tag{17}$$

The MAP estimate is computed using a numerical method. The texture parameters θ of the GGMRF model are estimated using the Minimum Mean Square Error (MMSE) estimation technique, and therefore given by a linear model as

$$\theta = \left(GG^T\right)^{-1}\left(G^T X\right) \tag{18}$$

where X are the observed coefficients inside the window of a size $N \times N$, and matrix G consists of neighboring coefficients attributed around each individual observed coefficient x_s inside a window of a size $N \times N$. The organization of a neighborhood for the bandelet and contourlet domain is shown in Fig. 9 and Fig. 10, respectively. Those figures show the parent-child relationships for the bandelet and contourlet transform. Each parameter θ weights the clique on different subbands.

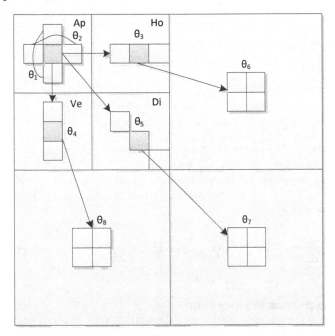

Fig. 9. Neighborhood cliques' organization for bandelet transforms

A logarithmic form can be introduced to simplify Eq. (16) as

$$\log p\left(y|x\right) \approx \sum_{i=1}^{N} \frac{1}{2}\left(\log\left(2\pi\right) - \log\left(h_{ii}\right)\right) + \log p\left(y_i|\hat{x}_i\right) + \log p\left(\hat{x}_i|\theta\right) \tag{19}$$

where h_{ii} are the diagonal elements of the Hessian matrix H, which has dimensions of $N \times N$, and N represents the dimension of moving window. Another approximation is then made

$$|H| \approx \prod_{i=1}^{N} h_{ii} \tag{20}$$

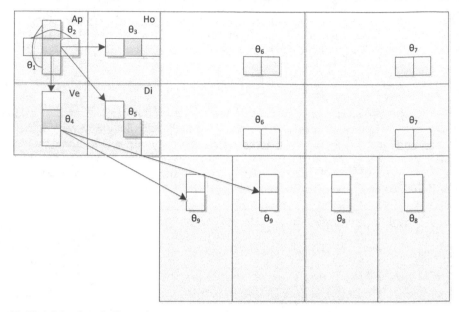

Fig. 10. Neighborhood cliques' organization for contourlet transforms

This approximation is possible, because all off-main diagonal elements represent covariances, and these are sparsely set matrixes that are close to zero, therefore those elements can be neglected. This assumption is in accordance with the statistical independence in Eq. (16). Only main-diagonal elements are needed for the Hessian matrix H, which are defined by

$$h_{ii} = \sum_{i=1}^{N} -v(v-1)\eta(v,\sigma_x)^2 \left(\eta(v,\sigma_x) \left| \hat{x}_i - \sum_{i \in \zeta} \theta_j \left(x_i^j + x_i^{j'} \right) \right| \right)^{v-2} - \sum_{i=1}^{N} \frac{\hat{x}_i}{\sigma_n^2} \qquad (21)$$

4. Outline of the proposed algorithm

1. The proposed despeckling algorithm transforms SAR images using bandelet (Le Pennec & Mallat, Apr 2005) or contourlet (da Cunha et al., 2006) transform. The number of decompositions depends on the size of the image. The number of levels l is chosen in such a way that the size of the approximation subband is larger than or equal to 64×64 pixels (minimum size).
2. The model parameters v and θ are estimated inside a window with a size of $N \times N$ pixels. In all experiments, a window with 7×7 pixels was used.
3. The noise and signal variances are estimated using (13) and (14), respectively.
4. The parameter θ is estimated using the MMSE defined in (18).
5. The shape parameter v is changed within the interval [0.5 ... 2.5] with a step size of 0.1.
6. The noise-reduced coefficients are estimated using the MAP estimate (15) for each value v. Each time, the texture parameters θ are estimated using the MAP estimate obtained in the previous step.

7.   The MAP estimate is used for the evidence estimation (21).
8.   The best MAP estimate $\hat{x}$ is accepted where the evidence has maximum value.
9.   The algorithm proceeds to the next pixel.

5. Experimental results

5.1 Synthetic SAR images

The synthetic SAR image, shown in Fig. 11, is composed of four different areas and with added four-look multiplicative speckle noise. The SAR image size shown in Fig. 11 is 512 × 512 pixels; therefore three levels of decomposition are used for bandelet transform. First let us show the difference between the pure bandelet and contourlet, and the MBD method.

Fig. 11. a) Original speckled image, b) Despeckled image obtained with the original bandelet denoising scheme, c) Despeckled image obtained with the original contourlet denoising scheme, and d) Despeckled image obtained with the MBD denoising technique

The bandelet transform is composed of a larger sliding window with a size of 16 × 16 (i.e. moving window per window), meanwhile inside a larger window, a smaller one with the size of 4 × 4 pixels moves on pixel basis. Those two sliding windows are used for searching the best decomposition inside the dyadic wavelet transform (Le Pennec & Mallat, Apr 2005). The contourlet transform is constructed using eight directions at the first level of decomposition. The last two levels are chosen to be dyadic, but this is not a requirement. The despeckled images obtained using the bandelet and contourlet transform are shown in Fig. 11 b) and c), respectively. Moreover, the despeckled image obtained using the MBD (Walessa & Datcu, 2000) is shown in Fig. 11 d).

And now with the MAP incorporated into the bandelet and contourlet transform.

Fig. 12. a) Original speckled image, b) Despeckled image obtained with the bandeleted denoising scheme, c) Despeckled image obtained with the contourlet denoising scheme, and d) Despeckled image obtained with the MBD denoising technique

$\mu = 127.94$	Bandelet	Contourlet	MBD
MSE	331	447	463
Mean	127.64	127.4	126.49
ENL ($\hat{x}$)	506	510	539
ENL (y/ $\hat{x}$)	3.14	3.2	3.18
Mean (y/ $\hat{x}$)	1.047	1.048	0.94

Table 1. Filter evaluation for synthetic test images

In Table 1, the objective measurements are presented for the denoising of image shown in Fig. 12. Objective measurements include the mean-square error (MSE), the equivalent number of looks (ENL), the mean value of the despeckled image, the ENL of speckle noise (ENL(y/ $\hat{x}$)), and the mean value of speckle noise y/ $\hat{x}$. The ENL of the image is given by μ^2/σ^2. The best MSE results are from bandelet transform in combination with Bayesian inference, thus having better results than those obtained from the contourlet transforms. A drawback of the contourlet transform is that it produces contours in the reconstructed image, which affects a MSE value. All wavelet-based methods well preserve the mean of the despeckled images. On the other hand, the MBD method well estimates the speckle noise, but it overblurs the reconstructed image, yielding a worse MSE value.

Figs. 13 a)-c) show the ratio images between the original and the reconstructed mosaic images obtained with bandelet, contourlet, and MDB. From these ratio images we can conclude that edges are well preserved, and that the speckle noise in the homogeneous areas is well estimated (i.e. removed) using the MBD method and second generation wavelets, as reported in Table 1.

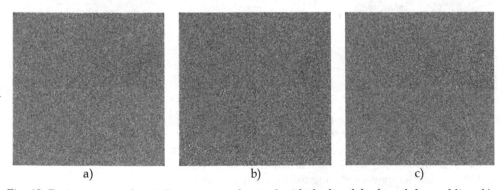

a)   b)   c)

Fig. 13. Ratio images y/ $\hat{x}$. a) Ratio image obtained with the bandelet-based despeckling, b) Ratio image obtained with the countourlet-based despeckling, and c) Ratio image obtained with the MBD method

The efficiency of the texture separation regarding the proposed method is demonstrated on four Brodatz textures, which are presented in a single mosaic composition and shown in Fig. 14. The textures are corrupted with a four-look speckle noise. The estimated parameter θ_2 obtained from bandelet and contourlet transforms is shown in Fig. 14 b) and c), respectively. The estimated texture parameters θ are classified into four classes using the K-means algorithm, and the classification results are shown in Fig. 14 d) and e), respectively. The best texture separation is obtained using a contourlet transform. The unsupervised classification of the texture parameters has an accuracy rate of 82 % and 89 %, for texture parameters obtained from the bandelet and contourlet transforms, respectively. Fig. 14 f) shows the classification of the texture parameter θ obtained with the MBD method. This method cannot well estimate classes on the right side of the image shown in Fig. 14 f).

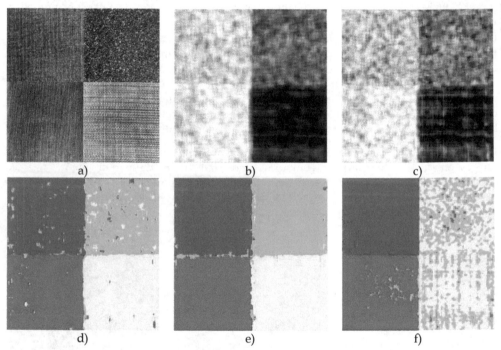

Fig. 14. a) Brodatz textures composed into a mosaic image, b) Texture parameter θ_2 obtained with the bandelet transform, c) Texture parameter θ_2 obtained with the contourlet transform, d) Classified parameter θ obtained with the bandelet transform using the K-means unsupervised classification into four classes, e) Classified parameter θ obtained with the contourlet transform using the K-means unsupervised classification into four classes, and f) Classified parameter θ obtained with the MBD method and K-means unsupervised classification into four classes

5.2 Real SAR images

The real SAR images are a sample images taken by TerraSAR-X satellite. The amplitude part of a single-look complex (SLC) SAR image is shown in Fig. 15 with a size of 2048×2048 pixels and ENL equal to 1.1. Four levels of dyadic decomposition are used for the bandelet decomposition. Five levels of contourlet transform are used, where the last two decompositions are dyadic and all other levels are contourlet directional subbunds consisted of eight directional subbands. The Daubechies symmetric four-filter bank (Daubechies, 1992) is used for the construction of bandelet and contourlet transforms.

a)

b)

c)

d)

Fig. 15. a) Original TerraSAR-X image © DLR (2007), b) Despeckled image obtained with the bandelet transform, c) Despeckled image obtained with the contourlet transforms, and d) Despeckled image obtained with the MBD method

Fig. 15 a)-d) shows the original SLC SAR image and the despeckled images obtained using the bandelet- and contourlet-based despeckling techniques, and the MBD despeckling method. Their ratio images are shown in Fig. 16 a)-c). The quality of the reconstructed images using the bandelet and contourlet transforms is nearly the same. However, the despeckling method based on bandelet transform has left out some speckle noise in the homogeneous regions. On the other hand, the homogeneous regions are well despeckled in the reconstructed image based on the contourlet transform, but undesired artifacts emerge in places around strong scatter returns in shape of lines, that are a consequence of contourlet subbands decomposition. This artifact is clearly visible in Fig. 16 b). Figs. 17 a)-c) show how strong scatterers are despeckled within the bandelet, contourlet, and dyadic wavelet domain.

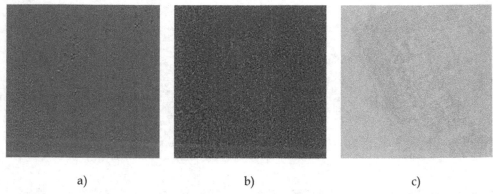

a)   b)   c)

Fig. 16. Ratio images y/ $\hat{x}$ for SAR images. a) Ratio image obtained with the bandelet transform, b) Ratio image obtained with contourlet transform, and c) Ratio image obtained with the MBD method

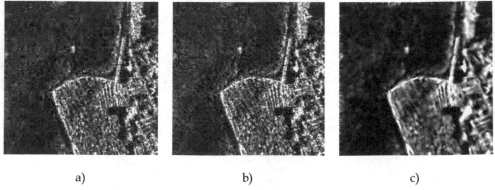

a)   b)   c)

Fig. 17. Despeckling of strong scatterers using a) bandelet transform, b) contourlet transform, and c) MBD method

The despeckling within the bandelet and dyadic wavelet domain are able to remove speckles around the strong scatterers, while the contourlet transform produces artifacts in this configuration. Higher image values are difficult to despeckle, because of the nature of the contourlet transform. Therefore, the noise is still present in those areas of the reconstructed image. However, the bandelet transform is overall computationally more demanding than contourlet transform (around 5.6 times), yet the despeckling of each contourlet subband takes about 4.5 times longer than with bandelet transform. Therefore, these methods are also computationally comparable.

To extract texture information from the denoised TerraSAR-X images we have used General Gauss-Markov Random Fields (GGMRF) as a prior pdf (Gleich & Datcu, 2007). As a prior pdf a first order model was used with cliques defined as Gauss-Markov Random Fields and shown in Figs. 9 and 10. Cliques were used to estimate central pixels for both transforms created in a 7×7 window which is moving throughout the whole picture. This was applied on transform's first approximation and its corresponding subbands. The texture parameters are iteratively estimated until second order Bayesian inference is increasing, which is used for finding the best model (Gleich & Datcu, 2007). The results of this method can be seen in Fig. 18, where the classification parameters for K-means algorithm were 5 classes and 7 iterations.

a)   b)   c)

Fig. 18. Comparison on information extraction. A) Original TerraSAR-X© image, b) Classified image on bandelet transform subbands, and c) Classified image on contourlet transform subbands

Texture parameters θ obtained during the despeckling procedure of the SAR image shown in Fig. 15 with bandelet, contourlet, and MBD method are shown in Fig. 19 b)-d). The algorithm used for classification into four different classes is the K-means algorithm. Fig. 19 a) is an indication of K-means algorithm applied to original image scene, where no textures can be identified as no processing was applied. The texture parameters obtained with both proposed algorithms clearly separate between homogeneous and heterogeneous areas. The contourlet transform compared to bandelet transform better separates the homogeneous and heterogeneous areas. From images it can be concluded, that contourlet transform is able to separate more heterogeneous areas from homogeneous ones. As a comparison, the MBD method can also distinct between homogeneous and heterogeneous areas as well as separate different textures in the scene, as shown in Fig. 19 d).

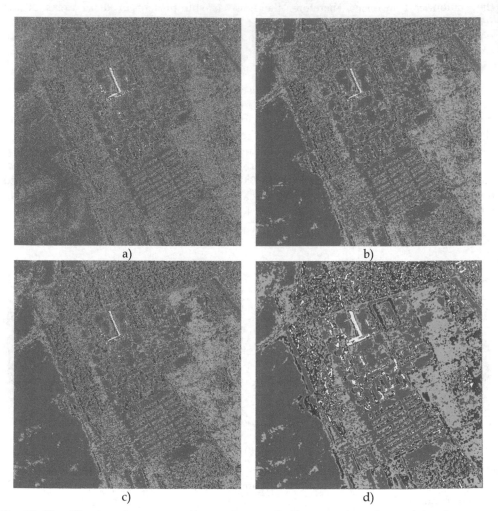

Fig. 19. Classification of texture parameter θ using the K-means algorithm and the a) original, b) bandelet, c) contourlet, and d) MBD-based algorithm

6. Conclusion

This book chapter has presented the proposed methods for despeckling a synthetic and real SAR images using second-generation wavelets. The Bayesian approach in incorporated into second generation wavelets using the wavelet domain. The prior and likelihood pdfs are modeled using GGMRF and Gaussian distribution. The second order Bayesian inference is used to better estimate model parameters and to find the best values possible. The evidence has been simplified and approximated using the Hessian approach. The experimental results have shown that the despeckling of real SAR images using second-generation wavelets is comparable with the dyadic wavelet-based despeckling algorithm (Gleich & Datcu, 2006). Moreover, information extracted using the contourlet domain gives good results using synthetic as well as real SAR data. Unfortunately, the contourlet-based despeckling introduces lines, which are consequences of cutting low-frequency components in the subband decomposition, which can be corrected by introducing a new filter or by post-processing step.

7. References

Argenti, F., and Alparone L. (2002). Speckle Removal From SAR Images in the Undecimated Wavelet Domain. *IEEE Tran. Geoscience and Remote Sensing*, Vol. 40, No. 11, (November 2002), pp. 2363-2374

Bamberger, R. H., and Smith, M. J. T. (1992). A filter bank for the directional decomposition of images: Theory and design. *IEEE Trans. Signal Proc.*, Vol. 40, No. 4, (April 1992), pp. 882-893

Chuna, A. L., Zhou, J. and Do, M. N. (2006). The Nonsubsampled Contourlet Transform: Theory, Design, and Application. *IEEE Tran. Image Processing*, Vol. 15, No. 10, (October 2006), pp. 3089-3101

Cui, J., and Wong W. (2006). Adaptive Chirplet Transform and Visual Evoked Potentials. *IEEE Tran. on Biomedical Engineering*, Vol. 53, No. 7, (July 2006), pp. 1378-1384

da Cunha, A. L., Zhou, J., and Do, M. N. (2006). The Nonsubsampled Contourlet Transform: Theory, Design, and Applications. *IEEE Transactions on Image Processing*, Vol. 15, No. 10, (October 2006), pp. 3089-3101

Dai, M., Cheng, P., Chan, A. K., and Loguinov D. (2004). Bayesian Wavelet Shrinkage with Edge Detection for SAR Image Despeckling. *IEEE Tran. Geoscience and Remote Sensing*, Vol. 42, No. 8, (August 2004), pp. 1642-1648

Daubechies, I. (June 1, 1992). *Ten Lectures on Wavelets* (1st edition), SIAM: Society for Industrial and Applied Mathematics, 978-0898712742, Philadelphia, Pennsylvania

Do, M. N. and Vetterli, M. (2005). The Contourlet transform: an efficient directional multiresolution image representation. *IEEE Trans. Image Process.*, Vol. 14, No. 12, (December 2005), pp. 2091-2106

Foucher, S., Benie, G. B., and Boucher, J. M. (2001). Multiscale MAP filtering of SAR Images. *IEEE Tran. Image Processing*, Vol. 10, No. 1, (January 2001), pp. 49-60

Frost, V. S., Stiles J. A., Shanmugan K. S., and Holtzman J. C. (1982). A model for radar images and its application to adaptive digital filtering of multiplicative noise. *IEEE Trans. Pattern Anal. Mach. Intell.*, Vol. 4, No. 2, (February 1982), pp. 157-166

Gleich, D., and Datcu, M. (2006). Gauss-Markov Model for SAR image Despeckling. *IEEE Signal Processing Letters.*, Vol. 13, No. 6, (June 2006), pp. 365-368

Gleich, D., and Datcu, M. (2007). Wavelet-Based Despeckling of SAR Images Using Gauss-Markov Random Fields. *IEEE Transactions on Geoscience and Remote Sensing*, Vol. 45, No. 12, (December 2007), pp. 4127–4143

Kuan, D. T., Sawchuk A. A., Strand T. C., and Chavel P. (1985). Adaptive noise smoothing filter for images with signal-dependent noise. *IEEE Trans. Pattern Anal. Mach. Intell.*, Vol. 7, No. 2, (February 1985), pp. 165-177

Le Pennec, E., and Mallat, S. (2001). Bandelet Image Approximation and Compression. In Proc. Int. Conf. Image Processing, ISBN: 0-7803-6725-1, Thessaloniki, Greece, October 2001

Le Pennec, E., and Mallat, S. (2005). Bandelet Image Approximation and Compression. *SIAM Journ. of Multiscale Modeling and Simulation*, Vol. 4, No. 3, (December 2005), pp. 992-1039

Le Pennec, E., and Mallat, S. (2005). Sparse geometric image representations with bandelets. *IEEE Transaction on Image Processing*, Vol. 14, No. 4, (April 2005), pp. 423–438

Lee, J. S. (1980). Digital image enhancement and noise filtering by use of local statistics. *IEEE Trans. Pattern Anal. Machine Intell.*, Vol. 2, No. 2, (March 1980), pp. 165-168

Li, Y., He, M. and Fang, X. (2006). A New Adaptive Algorithm for Despeckling of SAR images Based on Contourlet Transform. *IEEE Conference on Signal Processing*, ISBN 0-7803-9736-3, Guilin, China, November 2006

Peyré, G., and Mallat, S. (2005). Surface compression with geometric bandelets. *ACM Transactions on Graphics*, Vol. 24, No. 3, (July 2005), pp. 601-608

Sivia, D. S. (September 26, 1996). *Data Analysis: A Bayesian Tutorial* (1st edition), Oxford University Press, 978-0198518891, USA

Sveinsson, J. R., Semar Z., and Benediktsson, J. A. (2008). Speckle Reduction of SAR Images in the Bandelet Domain. *IEEE International Conference on Geoscience and Remote Sensing*, ISBN: 978-1-4244-2807-6, Boston, MA, July 2008

Vetterli, M. (1984). Multi-dimensional subband coding: some theory and algorithms. *Signal Processing*, Vol. 6, No. 2, (April 1984), pp. 97-112

Walessa, M., and Datcu M. (2000). Model-Based Despeckling and Information Extraction form SAR images. *IEEE Tran. Geoscience and Remote Sensing*, Vol. 38, No. 5, (September 2000), pp. 2258-2269

Wikipedia. (18 January 2011). TerraSAR-X, In: *Wikipedia*, October 6 2011, Available from: http://en.wikipedia.org/wiki/TerraSAR-X

Yang, S., Liu, F., Wand, M., and Jiao, L. (2007). Multiscale bandelet image compression. *International Symposium on Intelligent Signal Processing and Communication Systems*, ISBN 978-1-4244-1447-5, Xiamen, November 2007

4

Image Watermarking in Higher-Order Gradient Domain

Ehsan N. Arya, Z. Jane Wang and Rabab K. Ward
The University of British Columbia, Vancouver,
Canada

1. Introduction

With the widespread use of Internet and digital multimedia technologies, the interest in copyright protection of digital content has been rapidly increased. Digital watermarking has emerged as a possible solution for intellectual property rights protection. Watermarking has also proven to be a promising tool in many applications such as broadcast monitoring, fingerprinting, authentication and device control. In digital watermarking, additional information, called the *watermark*, is *imperceptibly* embedded into the original digital content.

Different applications pose different requirements on watermarking. For example, fragile watermarking is required in content authentication applications, while in applications such as copyright control the watermark should be robust to attacks[1]. In each application, the watermarking method makes a trade-off between the perceptual invisibility, robustness, security, data capacity and availability of side information. For instance, to increase the robustness of a watermark, the watermark strength needs to be increased, which in turn may make the watermark more visible. The invisibility requirement of watermarking limits the maximum amount of watermark bits (watermarking capacity) that can be embedded into a digital signal.

In the last two decades, a lot of work has been done in the field of image watermarking. The reader may refer to (Cox, 2008) for a survey of watermarking methods. Watermarking approaches can generally be classified into two categories (Wu & Liu, 2003): *spread spectrum* (SS) based watermarking (Cox et al., 1997; Podilchuk & Zeng, 1998) and *quantization* based watermarking (Chen & Wornell, 2001; Kundur & Hatzinakos, 2001; Moulin & Koetter, 2005). Below, these two approaches are discussed with some detail.

1.1 Spread spectrum watermarking

In general, any watermarking system that spreads the host signal over a wide frequency band can be called *spread spectrum* watermarking (Barni, 2003). In most SS type methods, a pseudo-random noise-like watermark is added (or multiplied) to the host feature sequence (Cox et al., 1997). While SS watermarking methods are robust to many types of attacks, they suffer from the host interference problem (Cox et al., 1999). This is because the host signal itself acts as a source of interference when extracting the watermark, and this may reduce the detector's performance.

---

[1] The attacks are defined as the processes that may impair the detection of the watermark.

The first approach to alleviate this problem is through designing better embedders at the encoder side. For example, the improved SS (ISS) method proposed in (Malvar & Florencio, 2003) exploits the information about the projection of the signal on the watermark. This knowledge is then used in the embedding process to compensate the signal interference.

Another approach to improve the performance of SS watermarking methods is to use the statistics of the host signal in the watermark detection (Zhong & Huang, 2006). Based on the distribution of the coefficients in the watermark domain, different types of optimum and locally optimum decoders have been proposed (Akhaee et al., 2010; Barni et al., 2003; Cheng & Huang, 2001; Hernandez et al., 2000; Kalantari et al., 2010).

1.2 Quantization-based watermarking

To overcome the host-interference problem, the quantization (random-binning-like) watermarking methods have been proposed. Chen and Wornell (Chen & Wornell, 2001) introduced *quantization index modulation* (QIM) as a computationally efficient class of data-hiding codes which uses the host signal state information to embed the watermark. In the QIM-based watermarking methods, a set of features extracted from the host signal are quantized so that each watermark bit is represented by a quantized feature value[2]. It has been shown that the QIM methods yield larger watermarking capacity than SS methods (Barni et al., 2003). The high capacity of these methods makes them more appropriate for data hiding applications.

Researchers have proposed different quantization-based watermarking methods. Gonzalez and Balado proposed quantized projection method that combines QIM and SS (Perez-Gonzlez & Balado, 2002). Chen and Lin (Chen & Lin, 2003) embedded the watermark by modulating the mean of a set of wavelet coefficients. Wang and Lin embedded the watermark by quantizing the super trees in the wavelet domain (Wang & Lin, 2004). Bao and Ma proposed a watermarking method by quantizing the singular values of the wavelet coefficients (Bao & Ma, 2005). Kalantari and Ahadi proposed a logarithmic quantization index modulation (LQIM) (Kalantari & Ahadi, 2010) that leads to more robust and less perceptible watermarks than the conventional QIM. Recently, a QIM-based method, that employs quad-tree decomposition to find the visually significant image regions, has also been proposed (Phadikar et al., 2011).

Since QIM methods do not suffer from the host-interference problem, their robustness to additive Gaussian noise is higher than that of SS methods. However, they are very sensitive to amplitude scaling attacks. Even small changes in the image brightness can significantly increase the bit error rate (BER) (Li & Cox, 2007). During the last few years, many improved techniques have been proposed to deal with this issue. These methods can be classified into the following main categories (Perez-Gonzalez et al., 2005):

- The first type of methods embed a pilot sequence in the signal (Eggers, Baeuml & Girod, 2002; Shterev & Lagendijk, 2005; 2006). Since the sequence is known to the decoder, it can be used to estimate any change in the signal amplitude.
- The second type of methods rely on designing amplitude-scale invariant codes, such as Trellis codes (Miller et al., 2004), orthogonal dirty paper codes (Abrardo & Barni, 2004) and order-preserving lattice codes (Bradley, 2004).

---

[2] The QIM method is discussed with more detail in subsection 2.5

- The third type of methods estimate the scaling factor based on the structure of the received data (Eggers, Bäuml & Girod, 2002; Lagendijk & Shterev, 2004; Shterev & Lagendijk, 2006).
- The fourth type of methods embed the watermark in the gain-invariant domains (Ourique et al., 2005; Perez-Gonzalez et al., 2005).

Among these methods, watermarking in a gain-invariant domain seems to be the best solution (Perez-Gonzalez et al., 2005).

1.3 The outline

The aim of this chapter is to describe how to insert robust, imperceptible and high-capacity watermark bits using a gain-invariant domain. Towards that goal we describe the *gradient direction watermarking* (GDWM) method (N. Arya et al., 2011). In this method, the watermark bits could be inserted into the angles of vectors of a higher-order gradient of the image, using the *angle quantization index modulation* (AQIM) method (Ourique et al., 2005). The AQIM has the advantages of QIM watermarking, but it also renders the watermark robustness to amplitude scaling attacks. In the GDWM method, the imperceptibility requirement is fulfilled by the following three mechanisms:

- By embedding the watermark in the significant (i.e. large) gradient vectors, the watermark becomes less perceptible. This is due to the observation that the human visual system (HVS) is less sensitive to distortions around the significant edges (i.e. represented by the significant gradient vectors) than to distortions in smooth areas (Barni et al., 2001).
- It is well known from comparing the additive with the multiplicative SS watermarking methods, that a disturbance proportional to the signal strength is more difficult to perceive (Langelaar et al., 2000). Therefore, by showing that a gradient change introduced by AQIM is proportional to the gradient magnitude, we can conclude that this method yields a less perceptible watermark.

  Assume that the angle of the gradient vector is altered by $\Delta\theta$. As the gradient vector g equals to $r\exp(i\theta)$, it is easy to obtain the absolute gradient change $|\Delta g|$ due to the angle change $\Delta\theta$:

$$|\Delta g| = |r\exp(i(\theta + \Delta\theta)) - r\exp(i\theta)| \approx r|\sin(\frac{\Delta\theta}{2})| \tag{1}$$

  where r denotes the gradient magnitude. It can be seen that the value of $|\Delta g|$ is proportional to r (i.e. $d \propto r$) and therefore AQIM results in a less perceptible watermark.

- The change in the higher order gradient vectors is less perceptible than the change in the first order gradients.

To increase the watermark capacity, the watermark bits are embedded in gradient vectors extracted from the multiscale wavelet coefficients of the image. This is accomplished by using a multiscale wavelet transform. For example, to embed a 256-bit watermark, 128, 64 and 64 bits can be embedded in the gradient fields obtained from scales 3, 4 and 5 of the wavelet transform of the image.

The rest of this chapter is organized as follows: a brief overview of the discrete wavelet transform (DWT), multi-scale gradient estimation, quantization index modulation (QIM) and angle quantization watermarking is given in Section 2. The watermark embedding scheme, called *gradient direction watermarking (GDWM)* (N. Arya et al., 2011), is described in Section 3. In this scheme, the image is first mapped to the wavelet domain from which the gradient fields are obtained. The gradient field at each wavelet scale is then partitioned into blocks. The watermark bits are embedded by changing the angles of the significant gradient vectors in each block using the AQIM method. The resultant (watermarked) wavelet coefficients of the image are computed. Finally the watermarked coefficients are inversely mapped to obtain the watermarked image. The decoding steps are discussed in Section 4, where the watermark bits are decoded following the reverse encoding steps, and the summary is given in Section 6.

2. Preliminaries

2.1 Notation

In this chapter, bold lower case letters, e.g. x, and bold capital letters, e.g. X, denote vectors (or discrete signals) and matrices, respectively. The vector or matrix elements are denoted by lower case letters with an index, e.g. x_i or x_{ij}. The vector at pixel (i, j) in the discrete vector field is represented as $f_{i,j}$.

2.2 Continuous wavelet transform (CWT)

Wavelet transform decomposes a signal into shifted and scaled versions of a mother wavelet. The *continuous wavelet transform (CWT)* (Mallat, 1997) of a 1-dimensional continuous signal x is defined by

$$W_{s,u}x = \langle x, \psi_{s,u} \rangle = \frac{1}{\sqrt{s}} \int_{-\infty}^{\infty} x(t)\psi\left(\frac{t-u}{s}\right) dt \qquad (2)$$

where ψ represents a *bandpass* wavelet function, called the *mother wavelet*, $\langle . \rangle$ denotes the inner product and the parameters s and u denote the *scale* and *translation (shift)*, respectively.

Similarly, the 2-dimensional *continuous wavelet transform (2D CWT)* of a continuous image $x(t_1, t_2)$ can be defined as

$$W_{s,u_1,u_2}x = \langle x, \psi_{s,u_1,u_2} \rangle = \frac{1}{\sqrt{s}} \int_{-\infty}^{\infty} \int_{-\infty}^{\infty} x(t_1, t_2)\psi\left(\frac{t_1-u_1, t_2-u_2}{s}\right) dt_1\, dt_2 \qquad (3)$$

where u_1 and u_2 denote the *horizontal* and *vertical* shifts, respectively.

2.3 Discrete Wavelet Transform (DWT)

Since the continuous wavelet transform is highly redundant, it is more efficient to sample the continuous shift-scale plane $u - s$, to obtain the *discrete wavelet transform* (DWT). In the dyadic DWT (i.e. $W_{j,n}x$), the scale and translation components, s and u, are respectively sampled at intervals of $\{2^j\}_{j\in\mathbb{Z}}$ and $\{n2^j\}_{n\in\mathbb{Z}}$, where j and n denote the discrete scale and translation parameters, respectively, and $\mathbb{Z}$ denotes the set of integer numbers.

The multiscale discrete wavelet transform represents a continuous signal $x(t)$ (or a discrete signal) in terms of bandpass filters $\psi_{j,n}(t)$ and shifted versions of a scaled lowpass filter $\phi(t)$ (called the *scaling function* (Mallat, 1997)) as

$$x(t) = \sum_{n\in\mathbb{Z}} a_J[n]\phi_{J,n}(t) + \sum_{j=1}^{J} \sum_{n\in\mathbb{Z}} d_j[n]\psi_{j,n}(t) \qquad (4)$$

where $a_J[n]$ and $d_j[n]$ are called *approximation* and *detail wavelet coefficients* at scales J and j, respectively. In Eq. (4), the wavelets $\psi_{j,n}(t)$ and $\phi_{J,n}(t)$ can be expressed as

$$\psi_{j,n}(t) = 2^{-j/2}\psi\left(2^{-j}t - n\right), \quad \phi_{J,n}(t) = 2^{-J/2}\phi\left(2^{-J}t - n\right) \tag{5}$$

For an *orthogonal* DWT The approximation and detail coefficients $a_J[n]$ and $d_j[n]$, are obtained by projecting $x(t)$ onto $\phi_{j,n}(t)$ and $\psi_{j,n}(t)$ as

$$a_J[n] = \left\langle x, \phi_{J,n} \right\rangle, \quad d_j[n] = \left\langle x, \psi_{j,n} \right\rangle. \tag{6}$$

DWT can also be extended to 2-dimensional images by decomposing the continuous image $x(t_1, t_2)$ using 2-dimensional scaling functions $\phi_{J,n_1,n_2}(t_1, t_2)$ and bandpass wavelets $\psi^k_{j,n_1,n_2}(t_1, t_2)$, such that

$$x(t_1, t_2) = \sum_{n_1, n_2 \in \mathbb{Z}} a_J[n_1, n_2]\phi_{J,n_1,n_2}(t_1, t_2) + \sum_{k=1}^{3}\sum_{j=1}^{J}\sum_{n_1, n_2 \in \mathbb{Z}} d_j^k[n_1, n_2]\psi^k_{j,n_1,n_2}(t_1, t_2), \tag{7}$$

As can be seen, at each scale j, the 2D DWT decomposes an image into 3 highpass subbands HL, LH and HH (denoted by superscript $k = 1, 2, 3$) and one lowpass subband LL. The approximation and detail wavelet coefficients in an orthogonal 2-dimensional DWT (2D DWT) can be obtained as

$$a_j[n_1, n_2] = \left\langle x, \phi_{j,n_1,n_2} \right\rangle, \quad d_j^k[n_1, n_2] = \left\langle x, \psi^k_{j,n_1,n_2} \right\rangle, \quad 1 \le k \le 3 \tag{8}$$

The 2-dimensional (2D) wavelets ϕ_{J,n_1,n_2} and ψ^k_{j,n_1,n_2} are usually obtained from tensor products of 1-dimensional (1D) orthogonal wavelets as

$$\phi_{J,n_1,n_2}(t_1, t_2) = \phi_{J,n_1}(t_1)\phi_{J,n_2}(t_2),$$
$$\psi^1_{j,n_1,n_2}(t_1, t_2) = \psi_{j,n_1}(t_1)\phi_{J,n_2}(t_2),$$
$$\psi^2_{j,n_1,n_2}(t_1, t_2) = \phi_{J,n_1}(t_1)\psi_{j,n_2}(t_2),$$
$$\psi^3_{j,n_1,n_2}(t_1, t_2) = \psi_{j,n_1}(t_1)\psi_{j,n_2}(t_2). \tag{9}$$

where $\psi^1_{j,n_1,n_2}(t_1, t_2)$, $\psi^2_{j,n_1,n_2}(t_1, t_2)$ and $\psi^3_{j,n_1,n_2}(t_1, t_2)$ are respectively the *horizontal, vertical* and *diagonal* continuous-time 2D wavelets that are shifted to the point (n_1, n_2).

Eqs. (3)-(9) can also be written in the discrete domain. For example, the 2-dimensional discrete time Haar wavelet transform uses the 1D discrete-time Haar lowpass vector $\boldsymbol{\varphi}_{1,0} = [+1, +1]^T$ and bandpass vector $\boldsymbol{\psi}_{1,0} = [-1, +1]^T$ to calculate the 2D discrete-time wavelets $\boldsymbol{\Psi}^1$, $\boldsymbol{\Psi}^2$ and $\boldsymbol{\Psi}^3$, as

$$\boldsymbol{\Psi}^1 = \boldsymbol{\psi}^T_{1,0} \otimes \boldsymbol{\varphi}_{1,0} = \begin{pmatrix} -1 & +1 \\ -1 & +1 \end{pmatrix},$$

$$\boldsymbol{\Psi}^2 = \boldsymbol{\varphi}^T_{1,0} \otimes \boldsymbol{\psi}_{1,0} = \begin{pmatrix} +1 & +1 \\ -1 & -1 \end{pmatrix},$$

$$\boldsymbol{\Psi}^3 = \boldsymbol{\psi}^T_{1,0} \otimes \boldsymbol{\psi}_{1,0} = \begin{pmatrix} -1 & +1 \\ +1 & -1 \end{pmatrix}. \tag{10}$$

where $\otimes$ and superscript T denote the *tensor product* and *matrix transpose*, respectively.

2.4 Multiscale gradient estimation Using 2D CWT

In this subsection, the relationship between the p-th order gradient vector of a continuous image and the 2-D wavelet coefficients is obtained.

Let us assume that the unshifted (i.e. $u_1 = u_2 = 0$) *horizontal* and *vertical* wavelets $\psi^1(t_1, t_2)$ and $\psi^2(t_1, t_2)$ have p vanishing moments, i.e.

$$\int_{-\infty}^{+\infty} t_1^k \, \psi^1(t_1, t_2) \, dt_1 = 0 \quad \text{and} \quad \int_{-\infty}^{+\infty} t_2^k \, \psi^2(t_1, t_2) \, dt_2 = 0 \quad \text{for} \ \ 0 \leq k \leq p. \tag{11}$$

It was shown in (Mallat & Hwang, 1992) that $\psi^1(t_1, t_2)$ and $\psi^2(t_1, t_2)$ with p vanishing moments can be written as

$$\psi^1(t_1, t_2) = (-1)^p \frac{\partial^p \rho(t_1, t_2)}{\partial t_1^p}, \quad \psi^2(t_1, t_2) = (-1)^p \frac{\partial^p \rho(t_1, t_2)}{\partial t_2^p} \tag{12}$$

where $\rho(t_1, t_2)$ is a smoothing function whose double integral is nonzero. Let $\psi_s^1(t_1, t_2)$ and $\psi_s^2(t_1, t_2)$ denote the scaled versions of $\psi^1(t_1, t_2)$ and $\psi^2(t_1, t_2)$, respectively, given as

$$\psi_s^1(t_1, t_2) = \frac{1}{s^2} \psi^1\left(\frac{t_1}{s}, \frac{t_2}{s}\right), \quad \psi_s^2(t_1, t_2) = \frac{1}{s^2} \psi^2\left(\frac{t_1}{s}, \frac{t_2}{s}\right) \tag{13}$$

where s denotes the continuous wavelet scale . Using Eqs. (12) and (13), it is easy to show that the horizontal and vertical wavelet components of the 2-D image $x(t_1, t_2)$ can be obtained as

$$W_s^1 x = x(t_1, t_2) * \psi_s^1(t_1, t_2) = (-s)^p \frac{\partial^p (x(t_1, t_2) * \rho_s(t_1, t_2))}{\partial t_1^p}$$

$$W_s^2 x = x(t_1, t_2) * \psi_s^2(t_1, t_2) = (-s)^p \frac{\partial^p (x(t_1, t_2) * \rho_s(t_1, t_2))}{\partial t_2^p} \tag{14}$$

where $*$ denotes the *convolution operator*, and

$$\rho_s(t_1, t_2) = \frac{1}{s^2} \rho\left(\frac{t_1}{s}, \frac{t_2}{s}\right) \quad \text{and} \quad W_s^1 x = \{W_{s,u_1,u_2}^1 x \mid \forall (u_1, u_2) \in \mathbb{R}^2\}. \tag{15}$$

where $W_{s,u_1,u_2}^1 x$ denotes the horizontal wavelet coefficient of x at scale s and point (u_1, u_2).

Eq. (14) shows that the vector of wavelet coefficients $W_{s,u_1,u_2} x = [W_{s,u_1,u_2}^1 x, W_{s,u_1,u_2}^2 x]^T$ can be interpreted as the pth-order gradient vector of $x * \rho_s$ at point (u_1, u_2).

2.5 Quantization Index Modulation (QIM)

To embed a watermark bit b in vector x, the QIM method quantizes x using the quantizer $\mathcal{Q}_0(.)$ when $b = 0$ and $\mathcal{Q}_1(.)$ when $b = 1$. The possible values of the quantizers $\mathcal{Q}_0(.)$ and $\mathcal{Q}_1(.)$ belong to the lattices Λ_0 and Λ_1, respectively:

$$\Lambda_0 = 2\Delta \mathbb{Z}^2$$
$$\Lambda_1 = 2\Delta \mathbb{Z}^2 + \Delta[1, 1]^T \tag{16}$$

where Δ and $\mathbb{Z}^2$ denote the *quantization step size* and the 2D set of integer values, respectively. for For the two-dimensional (2-D) QIM, the lattices Λ_0 and Λ_1 are shown in Fig. 1. The

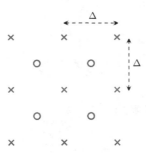

Fig. 1. Illustration of the 2-D uniform quantization index modulation (QIM). Lattices Λ_0 and Λ_1 in Eq. (16) are marked by $\times$ and $\circ$, respectively.

watermarked vector x^w is then obtained as the closest lattice point in Λ_0 to x when $b = 0$, or as the closest one in Λ_1 when $b = 1$. This can be expressed as

$$x^w = \begin{cases} \mathcal{Q}_0(x) & \text{if } b = 0 \\ \mathcal{Q}_1(x) & \text{if } b = 1 \end{cases} \tag{17}$$

At the decoder side, the value of the watermark bit $\hat{b}$ is extracted from the received vector x' by finding whether the nearest lattice point to the point x^w belongs to Λ_0 or Λ_1, i.e.

$$\hat{b} = \mathcal{B}(x') = \arg\min_{b \in \{0,1\}} \|x' - \mathcal{Q}_b(x')\|. \tag{18}$$

2.6 Angle quantization watermarking

As mentioned in subsection 1.2, the QIM method is fragile to amplitude scaling attacks. To address this concern, *angle quantization watermarking* embeds the watermark bit in the angle of the 2-dimensional vector. In angle quantization watermarking, the angle θ of vector x is assigned to a binary number 0 or 1.

In this section, we only describe the *uniform* angle quantization, in which the quantization circle is divided into a number of equiangular sectors in the range $(-\pi, \pi]$, as shown in Fig. 2. Two angle quantization watermarking methods are now described:

2.6.1 Angle Quantization Based Watermarking Method (AQWM)

The first method we consider is the angular version of the method proposed in (Kundur & Hatzinakos, 1999). The *quantization function*, denoted by $Q(\theta)$, maps a real angle θ to a binary number as follows:

$$Q(\theta) = \begin{cases} 0 & \text{if } \lfloor \theta/\Delta \rfloor \text{ is even} \\ 1 & \text{if } \lfloor \theta/\Delta \rfloor \text{ is odd} \end{cases} \tag{19}$$

where the positive real number Δ represents the *angular quantization step size* and $\lfloor . \rfloor$ denotes the floor function. To embed a watermark bit $b = 0$ or 1 into an angle $\theta \in [-\pi, \pi]$, the following rules are used:

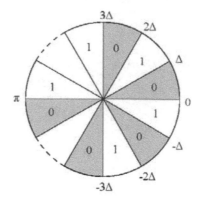

Fig. 2. The angle quantization circle with a fixed quantization step Δ.

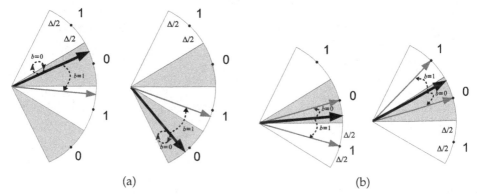

Fig. 3. Illustration of different angle quantization watermarking methods: (a) AQWM and (b) AQIM. Vectors before and after watermarking are represented by "thick black" and "thin gray" arrows, respectively.

- If $Q(\theta) = b$, then θ is not changed.
- If $Q(\theta) \neq b$, then θ is decreased by Δ if $\theta > 0$, and increased by Δ if $\theta \leq 0$.

These rules are illustrated in Fig. 3(a) and can be formulated as:

$$\theta^w = \begin{cases} \theta & \text{if } Q(\theta) = b \\ \theta - \Delta & \text{if } Q(\theta) \neq b \text{ and } \theta > 0 \\ \theta + \Delta & \text{if } Q(\theta) \neq b \text{ and } \theta \leq 0 \end{cases} \qquad (20)$$

where θ^w denotes the watermarked angle.

2.6.2 Angle Quantization Index Modulation (AQIM)

One drawback associated with AQWM is that if $Q(\theta) \neq b$, the angle is not necessarily modified toward the *nearest sector* having bit b. In other words, AQWM may change the angle more than required and this could lead to a perceptible watermark. Another drawback of AQWM is its low robustness to small angle perturbations. If the watermarked angle θ^w is

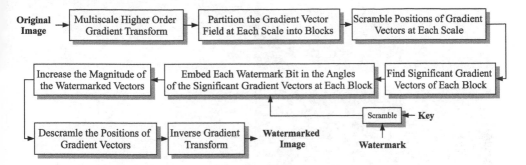

Fig. 4. The block diagram of the watermark embedding scheme.

close to sector boundaries, even a small amount of noise could be enough to make the angle pass the boundary and thus generate an error when the watermark bit is decoded.

To overcome these two drawbacks, the angle quantization index modulation (AQIM) method forms a possible alternative. AQIM (Ourique et al., 2005) is an extension of the QIM (Chen & Wornell, 2001) method, where the following rules are used to embed a watermark bit b into an angle θ:

- If $Q(\theta) = b$, then θ takes the value of the angle at the center of the sector it lies in.
- If $Q(\theta) \neq b$, then θ takes the value of the angle at the center of one of the two adjacent sectors, whichever is closer to θ.

These rules are shown in Fig. 3(b).

3. The gradient watermark embedding method

The block diagram of the gradient watermark embedding strategy (N. Arya et al., 2011) is shown in Fig. 4. The embedding steps are summarized as follows:

Step 1 To embed the watermark in the gradient of the image, first a domain that represents the gradient must be obtained. As shown in Eq. (14), the horizontal and vertical wavelet coefficients could be used to calculate the horizontal and vertical gradients of the image. Thus, based on the selected gradient order (e.g. 2nd, 3rd, 4th or 5th-order gradient), the image is transformed to the wavelet domain using the corresponding wavelet (e.g. Symlet2, Symlet3, Symlet4 or Symlet5). Thus, Symlet2 is used to obtain the 2nd-order gradient, Symlet3 to obtain the 3rd-order gradient and so on. The gradient field at a certain wavelet scale is then obtained using the the wavelet coefficients at the same scale. For example, (for gradient order 3) the gradient fields of image Lena at scales 3, 4 and 5 are obtained from the wavelet coefficients (obtained using Symlet3) at corresponding scales 3, 4 and 5, as shown in Fig. 5. The wavelet-based gradient estimation is described in subsection 3.1.

Step 2 To embed the bits of the watermark, the gradient field at each scale is partitioned into blocks (see Fig. 6(a)). The number of blocks depends on the number of bits to be embedded. Thus, bits can be embedded in the gradient fields corresponding to more than one scale.

Step 3 The positions of the gradient vectors are uniformly scrambled at each scale, as illustrated in Fig. 6(b). The watermark bits are inserted into the significant gradient

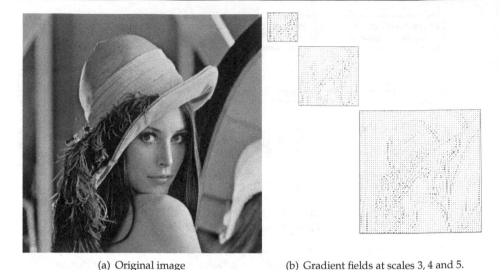

(a) Original image   (b) Gradient fields at scales 3, 4 and 5.

Fig. 5. Image Lena and its representations in the gradient domain. The Symlet3 wavelet is used to obtain the gradient vectors.

vectors of each block. Significant gradient vectors are the gradient vectors with large magnitudes. Embedding the watermark bits in the significant vectors makes it robust to many attacks. As some blocks do not contain significant gradients, and as a watermark bit is inserted into each block, scrambling the locations of the significant gradient vectors is used. The scrambling used should guarantee that statistically each block contains at least one significant gradient vector. More details about the scrambling method are given in subsection 3.2.

Step 4 The significant gradient vectors of each block are calculated.

Step 5 For security reasons, the binary watermark message is scrambled using a secret key.

Step 6 In each block, one bit of the watermark is embedded in the angle of the most significant gradient vectors, using angle quantization index modulation (AQIM) (see Fig. 6(c)). It is preferred however to embed the same bit using 2 (or even more) gradient vectors. The number of the most significant gradient vectors as the bit is embedded in is denoted by BR. The AQIM method in the gradient domain is discussed in subsection 3.3.

Step 7 The correct detectability of the watermarked gradient vectors is enhanced by increasing their magnitudes relative to the nonsignificant (unwatermarked) vectors, as illustrated in Fig. 6(d). More explanations are given in subsection 3.4.

Step 8 The watermarked gradient fields at each scale are descrambled, using the descrambling method associated with the scrambling method in step 3 (cf. Fig. 6(e)). The descrambling method is explained in subsection 3.2.

Step 9 The watermarked wavelet coefficients are obtained from the watermarked gradient vectors.

Step 10 Finally, the watermarked image is obtained using the inverse wavelet transform on the watermarked wavelet coefficients.

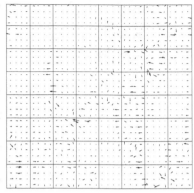

(a) Partitioned gradient field at scale 4.

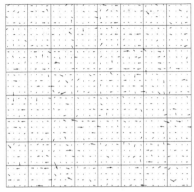

(b) Scrambled gradient field

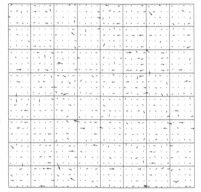

(c) Watermarked scrambled gradient field

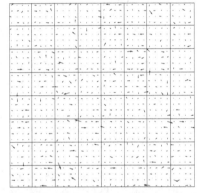

(d) Enhanced watermarked gradient field

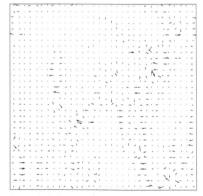

(e) Descrambled watermarked gradient field

(f) Watermarked image of Lena

Fig. 6. Illustration of different steps during embedding a pseudo-random binary watermark of size 8 × 8 into image Lena. A 64-bit watermark is inserted into the gradient field obtained using Symlet3 at scale 4.

3.1 Higher-order multiscale gradient transform

To obtain the gradient vector $g_j[m] = g_{hj}[m] + i\, g_{vj}[m]$ at pixel $m = (m_1, m_2)$ and wavelet scale j, the horizontal and vertical wavelet coefficients $d_j^1[m]$ and $d_j^2[m]$ are used:

$$\begin{pmatrix} g_{hj}[m] \\ g_{vj}[m] \end{pmatrix} = A_j \begin{pmatrix} d_j^1[m] \\ d_j^2[m] \end{pmatrix} \tag{21}$$

and using Eq. (14) matrix A_j is obtained as

$$A_j = \begin{pmatrix} 1 & 0 \\ 0 & 1 \end{pmatrix} \tag{22}$$

Based on the type of the gradient operator used to calculate the gradient vectors, other versions of matrix A_j can also be used (N. Arya et al., 2011). In this chapter, the identity matrix is used to obtain the higher-order gradient vectors at each scale. This means that the horizontal and vertical wavelet coefficients are the same as the gradient vector components.

3.2 Scrambling and descrambling the gradient fields

As mentioned before, to embed the watermark bits, the gradient fields of the image are obtained. The straightforward way to embed the watermark bits is to partition the gradient fields into non-overlapping blocks and each watermark bit is then embedded into each block. The bit is inserted into the BR most significant gradient vectors of the block, since embedding the watermark in the significant vectors makes it robust to attacks. However in natural images, the spatial distribution of the significant gradient vectors in the gradient fields is non-uniform (as some parts of the image may have all or most of the significant gradient vectors, while other parts may have no significant gradient vectors). If a bit is embedded into a block with no significant gradient vectors, then the robustness of the watermark bit to noise and other attacks is reduced. Therefore the straightforward uniform embedding may reduce the robustness of the watermark bits in the image areas with no significant gradient vectors.

To solve this problem, the locations of all gradient vectors are uniformly scrambled, so that each block contains at least one significant gradient vector. As shown in Fig. 6(c), the positions of the gradient vectors at each scale are uniformly scrambled over the gradient field. At each block, the watermark bits are inserted into the angles of significant gradient vectors, using AQIM method. The gradient vectors are then descrambled so they are located back at their original positions.

3.2.1 Scrambling method

The scrambling method should be a geometric transform that would ideally result in a uniform distribution of the locations of the significant gradient vectors. Different geometric image scrambling methods have been proposed. These include the Fibonacci transformation (Zou et al., 2004), Arnold Cat transformation (Ming & Xi-jian, 2006) and Gray Code transformation (Zou & Ward, 2003). It has been shown in (Xue Yang & Jia, 2010) that most of the geometric transforms are special cases of the *affine modular transformation* (Zou et al., 2005). Therefore, the affine modular transform is employed, due to its generality.

In a gradient field of size $M \times M$, let the coordinate vectors of each vector in the original and scrambled gradient fields be denoted by $[m_1, m_2]^T$ and $[m'_1, m'_2]^T$, respectively. The *affine modular mapping* from $[m_1, m_2]^T$ to $[m'_1, m'_2]^T$ is defined as

$$\begin{pmatrix} m'_1 \\ m'_2 \end{pmatrix} = \left[\begin{pmatrix} a & b \\ c & d \end{pmatrix} \begin{pmatrix} m_1 \\ m_2 \end{pmatrix} + \begin{pmatrix} e \\ f \end{pmatrix} \right] \mathrm{mod}(M) \tag{23}$$

where a, b, c, d, e and f are scrambling parameters. If the absolute value of the determinant of the matrix $S = [a, b; c, d]$ equals to 1 (i.e. $det(S) = |ad - bc| = 1$), the transform is area preserving and one-to-one mapping. For a given positive integer M, the *necessary* and *sufficient* condition for the *periodicity* of the affine modular transform is that $det(S)$ and M are primal to each other (Zou et al., 2005).

3.2.2 Descrambling method

By using the periodicity property of the affine modular transform, the original image can be recovered from the scrambled image. Let the smallest period of this map be denoted by t_M. If the scrambled image is constructed by applying this transformation t times, the descrambled image can be exactly recovered after consecutively applying the same transformation $t_M - t$ times (Xue Yang & Jia, 2010). To reduce the number of parameters needed in embedding and to make it an area preserving transform, the elements of the mapping matrix S are assigned the values $a = 1$, $b = p$, $c = q$ and $d = 1 + pq$. The mapping matrix S is then given by

$$S = \begin{pmatrix} 1 & p \\ q & 1 + pq \end{pmatrix}. \tag{24}$$

3.3 AQIM in the gradient domain

As mentioned before, to insert the watermark bits, the angles of the significant gradient vectors are changed. The amount of change $\Delta\theta$ in the angle is obtained by Eq. (??). To rotate a gradient vector by $\Delta\theta = \theta^w - \theta$, the *rotation matrix* $R_{\Delta\theta}$ (of size 2×2) is used:

$$R_{\Delta\theta} = \begin{pmatrix} \cos(\Delta\theta) & -\sin(\Delta\theta) \\ \sin(\Delta\theta) & \cos(\Delta\theta) \end{pmatrix}. \tag{25}$$

Thus, the watermarked gradient is obtained as

$$g_j^w[m] = R_{\Delta\theta} g_j[m] \tag{26}$$

3.4 Enhancing the detectability of the watermarked gradient vectors

Attacks may change the watermark bits in one of the following ways:

- The attack may result in a change in the angle of a significant gradient vector, such that the extracted bit is different from the embedded one.
- The attack may result in a reduction in the magnitude of a significant gradient vector, such that the decoder can no longer identify it as a significant vector. In this case, a watermark bit is extracted from a gradient vector that was not amongst the BR most significant vectors.

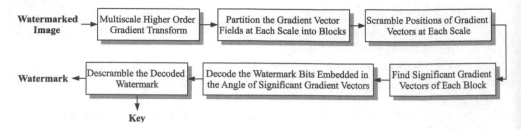

Fig. 7. The block diagram of the watermark decoding scheme.

In the first case, to increase the watermark robustness, a larger angle quantization step size Δ is preferred. Note that the maximum step size is constrained by the imperceptibility constraint on the embedded watermark. In the second case, one solution is to enhance the detectability of the watermarked gradient vectors by increasing the magnitude r of such a vector.

Let us denote the magnitude of the gradient vector g_j by r. To increase the magnitude of each watermarked vector, dr can be calculated by:

$$dr = \frac{\alpha}{r} e^{-(r - r_{LIS})} \tag{27}$$

where α is a constant that adjusts the overall gradient magnitude change in the image and r_{LIS} is the magnitude of the *largest insignificant* gradient vector in each block.

4. The watermark decoding method

The block diagram of the watermark decoding method is shown in Fig. 7. The decoding steps are as follows:

Step 1 Step 1 of watermark embedding process is repeated. That is the watermarked image is mapped to the gradient domain using the multiscale gradient transform, discussed in subsection 3.1.

Step 2 As in step 2 of the watermark embedding, the gradient field at each scale is partitioned into blocks. The blocks should be of the same size as those used in the embedding process.

Step 3 The positions of the gradient vectors are uniformly scrambled with the same method and parameters used in step 3 of the the embedding process.

Step 4 The significant gradient vectors of each block are detected.

Step 5 The bits of the BR most significant gradient vectors of each block are decoded, using the AQIM decoder, as discussed in subsection 2.6. If all the BR decoded bits have the same value (e.g. $\hat{b}$), the decoded watermark bit will also be the same bit (i.e. $\hat{b}$). In the case, the bits have different values, they are assigned weights based on the following rules:

- A watermark bit extracted from a large gradient vector is given more weight than a bit extracted from a small gradient vector.
- A watermark bit extracted from an angle close to the center of a sector, is given more weight than a bit extracted from an angle close to a sector boundary.

Based on these two rules, each watermark bit is weighted by

$$a_k = r^\gamma \cdot \left[\frac{\Delta}{2} - \left| |\theta| - \left(\Delta \lfloor \frac{|\theta|}{\Delta} \rfloor + \Delta/2 \right) \right| \right], \ k = 1, \ldots, BR \tag{28}$$

Image	Δ_3 (rad)	Δ_4 (rad)	Δ_5 (rad)	α	PSNR (dB)
Lena	$\pi/10$	$\pi/22$	$\pi/24$	0.05	42 dB
Barbara	$\pi/12$	$\pi/20$	$\pi/26$	0.5	42 dB
Baboon	$\pi/12$	$\pi/16$	$\pi/18$	0.60	42 dB
Peppers	$\pi/12$	$\pi/22$	$\pi/26$	0.25	42 dB

Table 1. Values of the quantization step size Δ, the gradient magnitude enhancement coefficient α, and the PSNR of the watermarked image.

where BR denotes the number of watermarked significant gradient vectors in each block, γ is a constant that represents the importance of magnitude weighting vs. angle weighting. Based on the weights determined by Eq. (28), the watermark bit in each block after decoding is given the value

$$\hat{b} = \begin{cases} 1 \text{ if } b^w \geq 0.5 \\ 0 \text{ Otherwise} \end{cases} \tag{29}$$

where b^w is defined as

$$b^w = \frac{\sum\limits_{k=1}^{BR} a_k \mathcal{B}(\theta_k)}{\sum\limits_{k=1}^{BR} a_k} \tag{30}$$

where the function $\mathcal{B}(\theta)$ is as defined in Eq. (18).

Step 6 Since the extracted watermark is the scrambled version of the original message, it should be descrambled using the same key used in the embedding process.

5. Example

Different pseudo-random binary watermarks of size 256 are embedded in the grayscale images *Lena*, *Barbara*, *Baboon* and *Peppers*. All the images are of size 512×512. To obtain the 2nd, 3rd, 4th and 5th order gradients, Symlet2, Symlet3, Symlet4 and Symlet5 wavelets are used, respectively. The Symlet wavelets are chosen because of their near-symmetrical shapes. Irrespective of the order of the gradient used, for a 256-bit watermark, 128, 64 and 64 bits are embedded in the gradient fields at scales 3, 4 and 5 using the block sizes 4×8, 4×4 and 2×2, respectively.

The parameters (p, q) for images Pappers, Baboon, Barbara and Lena are set to $(3, 4)$, $(1, 1)$, $(2, 3)$ and $(2, 1)$, respectively. The value of the parameter γ in Eq. (28) is set to 4. Each bit is embedded in the 2 most significant (largest) gradient vectors of each block, i.e. BR=2. The angular quantization step size Δ and the gradient magnitude enhancement coefficient α, given in Eq. (27), are obtained separately for each image and gradient transform. Table **??** shows the optimum values of Δ for each image. To evaluate the robustness of the scheme, each watermarked image is distorted by different types of attacks. After the attacks, each watermark is extracted and is compared with the original watermark to estimate the *bit error rate* (BER). The overall BER is obtained by averaging over 100 runs with 100 different pseudo-random binary watermarks.

The BER (%) results of the *gradient direction watermarking* (GDWM) method, under amplitude scaling, Gaussian filtering, median filtering and JPEG compression is shown in Table **??**. It can be seen that GDWM is robust to amplitude scaling attack, no matter which gradient transform

Image	Wavelet	Amplitude Scaling (scale=2)	Gaussian Filter		Median Filter		JPEG (Q)	
			7×7	9×9	3×3	5×5	10	20
Lena	Sym2	0	1.19	4.99	0.66	5.23	8.17	0.45
Lena	Sym3	0	0	3.23	0.33	4.61	7.79	0.48
Lena	Sym4	0	6.36	15.78	0.51	9.42	16.95	4.20
Lena	Sym5	0	**0**	**0.95**	**0**	**2.83**	**6.67**	**0.32**
Barbara	Sym2	0	0.64	2.20	0	3.46	7.57	0.49
Barbara	Sym3	0	0.59	1.74	**0**	2.57	7.35	**0.28**
Barbara	Sym4	0	3.10	11.84	1.57	16.50	19.20	5.55
Barbara	Sym5	0	**0.09**	**1.22**	0.41	**2.41**	**6.62**	0.32
Baboon	Sym2	0	0.94	5.95	2.27	14.81	**6.92**	0.16
Baboon	Sym3	0	1.23	**4.65**	1.91	18.33	7.59	**0.13**
Baboon	Sym4	0	4.40	16.26	7.29	28.16	18.83	5.99
Baboon	Sym5	0	**0.18**	5.70	**0.98**	**14.80**	8.15	0.15
Peppers	Sym2	0	0.93	5.66	0.51	4.43	7.72	0.79
Peppers	Sym3	0	2.28	6.59	**0.01**	3.81	7.69	**0.36**
Peppers	Sym4	0	24.97	37.16	1.07	10.95	21.13	5.91
Peppers	Sym5	0	**0.92**	**3.25**	0.22	**2.81**	**7.50**	0.48

Table 2. The BER (%) results of GDWM under different types of attacks.

is used. For the Gaussian filtering, median filtering and JPEG compression attacks, Symlet5 (i.e. the 5th-order gradient of the image) gives the best results.

Fig. 8 presents the image Lena watermarked with the GDWM method, using Symlet2, Symlet3, Symlet4 and Symlet5 wavelets. To compare the original and watermarked images, the SSIM metric is employed due to its compatibility with the human visual system (Wang et al., 2004). As shown in Fig. 8, the GDWM method yields imperceptible watermarks at PSNR=42 dB. Based on visual inspection, Symlet4 (i.e. the gradient of order 4) gives the most imperceptible watermark. However, the best SSIM value is obtained by Symlet5 (i.e. the gradient of order 5).

Table 3 compares the BER results of the GDWM method with the method in (Wang et al., 2002) under median filtering, JPEG compression, AWGN and salt & pepper noise attacks. As in (Wang et al., 2002), the watermark length in both methods is 256 bits and the PSNR of all the watermarked test images is 42 dB. Symlet5 wavelet is used to implement the GDWM method.

Image	Method	Median Filter 3×3	JPEG $Q = 11$	AWGN $\sigma = 10$	S&P $p = 0.01$
Lena	Wang	30.80	29.80	1.45	2.45
Lena	GDWM	**0**	**5.54**	**0.29**	**1.67**
Barbara	Wang	24.95	16.45	1.45	2.25
Barbara	GDWM	**0.41**	**4.73**	**0.16**	**2.12**
Baboon	Wang	31.65	16.95	1.30	1.95
Baboon	GDWM	**0.98**	**4.61**	**0.11**	**1.60**
Peppers	Wang	29.35	26.10	1.25	**2.00**
Peppers	GDWM	**0.22**	**5.67**	**0.40**	2.60

Table 3. The BER comparisons between the GDWM and Wang's method (Wang et al., 2002) under different types of attacks (Message length=256 bits, PSNR=42 dB)

(a) Symlet2, SSIM=0.9944   (b) Symlet3, SSIM=0.9935

(c) Symlet4, SSIM=0.9958   (d) Symlet5, SSIM=0.9963

Fig. 8. Test image Lena watermarked using different gradient transforms in the GDWM method. In each case, a 256-bit watermark is embedded with PSNR=42dB.

The results clearly demonstrate that the GDWM method consistently outperforms the Wang's method under all considered attacks.

In Table 4 the GDWM is compared with the non-blind method proposed in (Akhaee et al., 2010). In both methods, a 128-bit pseudo-random binary message is embedded in the images Baboon and Barbara with PSNRs 39.53 dB and 36.63 dB, respectively. To embed 128 bits, 64, 32 and 32 bits are inserted in the gradient fields at scales 3, 4 and 5, respectively. Symlet5 wavelet is used to obtain the 5th-order gradient fields. It can be seen that the GDWM method outperforms Akhaee's method (Akhaee et al., 2010) under the spatial scaling, JPEG compression and AWGN noise attacks. However, GDWM is slightly less robust under the salt & pepper noise attacks.

Image	Method	Scaling $(s = 0.8)$	JPEG $(Q = 20)$	AWGN $(\sigma = 20)$	S&P $(p = 0.05)$
Barbara	Akhaee	2.34	0.40	0.10	**1.48**
Barbara	GDWM	**0**	**0**	**0**	1.61
Baboon	Akhaee	3.20	1.80	0.30	**2.89**
Baboon	GDWM	**0**	**0.22**	**0.16**	4.31

Table 4. The BER comparison between the GDWM method and Akhaee's method (Akhaee et al., 2010) under different types of attacks (Message length=128 bits)

6. Conclusion

This chapter describes a gradient-based image watermarking method, called gradient direction watermarking (GDWM). In this method, the watermark bits are embedded in the angles of the gradient vectors of the image. The gradient vectors correspond to a higher-order gradient of the image. The gradient fields are obtained from the wavelet coefficients of the image at different scales. To embed the watermark bit in the gradient angle, the angle quantization index modulation (AQIM) method is used. AQIM makes the watermark both imperceptible and robust to amplitude scaling attack.

The GDWM method is tested on different real images. The experimental results show that implementing the GDWM method in the 5th-order gradient domain (obtained using using Symlet5 wavelet) yields both robust and imperceptible watermarks. It is also shown that the GDWM outperforms other watermarking methods and it is robust to a wide range of attacks.

7. References

Abrardo, A. & Barni, M. (2004). Orthogonal dirty paper coding for informed data hiding, *Security, Steganography, and Watermarking of Multimedia Contents VI* 5306(1): 274–285.

Akhaee, M., Sahraeian, S. & Marvasti, F. (2010). Contourlet-based image watermarking using optimum detector in a noisy environment, *IEEE Trans. on Image Proces.* 19(4): 967 –980.

Bao, P. & Ma, X. (2005). Image adaptive watermarking using wavelet domain singular value decomposition, *IEEE Trans. on Circ. and Sys. for Video Tech.* 15(1): 96 – 102.

Barni, M. (2003). What is the future for watermarking? (part ii), *Signal Processing Magazine, IEEE* 20(6): 53 – 59.

Barni, M., Bartolini, F., De Rosa, A. & Piva, A. (2003). Optimum decoding and detection of multiplicative watermarks, *IEEE Trans. on Signal Proces.* 51(4): 1118 – 1123.

Barni, M., Bartolini, F. & Piva, A. (2001). Improved wavelet-based watermarking through pixel-wise masking, *IEEE Trans. on Image Proces.* 10(5): 783 –791.

Bradley, B. A. (2004). Improvement to cdf grounded lattice codes, *Security, Steganography, and Watermarking of Multimedia Contents VI* 5306(1): 212–223.

Chen, B. & Wornell, G. (2001). Quantization index modulation: a class of provably good methods for digital watermarking and information embedding, *IEEE Trans. on Information Theory* 47(4): 1423 –1443.

Chen, L.-H. & Lin, J.-J. (2003). Mean quantization based image watermarking, *Image and Vision Computing* 21(8): 717 – 727.

Cheng, Q. & Huang, T. (2001). An additive approach to transform-domain information hiding and optimum detection structure, *IEEE Trans. on Multimedia* 3(3): 273 –284.

Cox, I. (2008). *Digital watermarking and steganography*, The Morgan Kaufmann series in multimedia information and systems, Morgan Kaufmann Publishers.

Cox, I., Kilian, J., Leighton, F. & Shamoon, T. (1997). Secure spread spectrum watermarking for multimedia, *IEEE Trans. on Image Proces.* 6(12): 1673 –1687.

Cox, I., Miller, M. & McKellips, A. (1999). Watermarking as communications with side information, *Proceedings of the IEEE* 87(7): 1127 –1141.

Eggers, J. J., Baeuml, R. & Girod, B. (2002). Estimation of amplitude modifications before scs watermark detection, *Security and Watermarking of Multimedia Contents IV* 4675(1): 387–398.

Eggers, J. J., Bäuml, R. & Girod, B. (2002). Estimation Of Amplitude Modifications before SCS Watermark Detection, *Proceedings of SPIE Security and Watermarking of Multimedia Contents IV* 4675: 387–398.

Hernandez, J., Amado, M. & Perez-Gonzalez, F. (2000). Dct-domain watermarking techniques for still images: detector performance analysis and a new structure, *Image Processing, IEEE Transactions on* 9(1): 55 –68.

Kalantari, N. & Ahadi, S. (2010). A logarithmic quantization index modulation for perceptually better data hiding, *IEEE Trans. on Image Proces.* 19(6): 1504 –1517.

Kalantari, N., Ahadi, S. & Vafadust, M. (2010). A robust image watermarking in the ridgelet domain using universally optimum decoder, *IEEE Trans. on Circ. and Sys. for Video Tech.* 20(3): 396 –406.

Kundur, D. & Hatzinakos, D. (1999). Digital watermarking for telltale tamper proofing and authentication, *Proc. of the IEEE* 87(7): 1167 –1180.

Kundur, D. & Hatzinakos, D. (2001). Diversity and attack characterization for improved robust watermarking, *IEEE Trans. on Signal Proces.* 49(10): 2383 –2396.

Lagendijk, R. & Shterev, I. (2004). Estimation of attacker's scale and noise variance for qim-dc watermark embedding, 1: 55–58.

Langelaar, G., Setyawan, I. & Lagendijk, R. (2000). Watermarking digital image and video data. a state-of-the-art overview, *IEEE Signal Proces. Mag.* 17(5): 20 –46.

Li, Q. & Cox, I. (2007). Using perceptual models to improve fidelity and provide resistance to valumetric scaling for quantization index modulation watermarking, *IEEE Transactions on Information Forensics and Security* 2(2): 127 –139.

Mallat, S. (1997). *A Wavelet Tour of Signal Proces.*, AP Professional, London.

Mallat, S. & Hwang, W. (1992). Singularity detection and processing with wavelets, *Information Theory, IEEE Transactions on* 38(2): 617 –643.

Malvar, H. & Florencio, D. (2003). Improved spread spectrum: a new modulation technique for robust watermarking, *IEEE Transactions on Signal Processing* 51(4): 898 – 905.

Miller, M., Doerr, G. & Cox, I. (2004). Applying informed coding and embedding to design a robust high-capacity watermark, *IEEE Trans. on Image Proces.* 13(6): 792 –807.

Ming, C. & Xi-jian, P. (2006). Image steganography based on arnold transform, *Computer appl. research* 1: 235–237.

Moulin, P. & Koetter, R. (2005). Data-hiding codes, *Proc. of the IEEE* 93(12): 2083 –2126.

N. Arya, E., Wang, Z. & Ward, R. (2011). Robust image watermarking based on multiscale gradient direction quantization, *Information Forensics and Security, IEEE Transactions on* (99): 1.

Ourique, F., Licks, V., Jordan, R. & Perez-Gonzalez, F. (2005). Angle qim: a novel watermark embedding scheme robust against amplitude scaling distortions, *Proc. IEEE Int. Conf. on Acoust., Speech, and Signal Proces. (ICASSP '05)* 2: ii/797 – ii/800 Vol. 2.

Perez-Gonzalez, F., Mosquera, C., Barni, M. & Abrardo, A. (2005). Rational dither modulation: a high-rate data-hiding method invariant to gain attacks, *IEEE Trans. on Signal Proces.* 53(10): 3960 – 3975.

Perez-Gonzlez, F. & Balado, F. (2002). Quantized projection data hiding, *Proc. of Int. Conf. on Image Proces.* 2: 889–892.

Phadikar, A., Maity, S. P. & Verma, B. (2011). Region based QIM digital watermarking scheme for image database in DCT domain, *Computers & Electrical Engineering* In Press.

Podilchuk, C. & Zeng, W. (1998). Image-adaptive watermarking using visual models, *IEEE Journal on Sel. Areas in Comm.* 16(4): 525 –539.

Shterev, I. D. & Lagendijk, R. L. (2005). Maximum likelihood amplitude scale estimation for quantization-based watermarking in the presence of dither, *Security, Steganography, and Watermarking of Multimedia Contents VII* 5681(1): 516–527.

Shterev, I. & Lagendijk, R. (2006). Amplitude scale estimation for quantization-based watermarking, *Signal Processing, IEEE Transactions on* 54(11): 4146 –4155.

Wang, S.-H. & Lin, Y.-P. (2004). Wavelet tree quantization for copyright protection watermarking, *IEEE Trans. on Image Proces.* 13(2): 154 –165.

Wang, Y., Doherty, J. & Van Dyck, R. (2002). A wavelet-based watermarking algorithm for ownership verification of digital images, *IEEE Trans. on Image Proces.* 11(2): 77 –88.

Wang, Z., Bovik, A., Sheikh, H. & Simoncelli, E. (2004). Image quality assessment: from error visibility to structural similarity, *IEEE Transactions on Image Processing* 13(4): 600 –612.

Wu, M. & Liu, B. (2003). Data hiding in image and video .i. fundamental issues and solutions, *IEEE Trans. on Image Proces.* 12(6): 685 – 695.

Xue Yang, Xiaoyang Yu, Q. Z. & Jia, J. (2010). Image encryption algorithm based on universal modular transformation, *Inf. Tech. Journal* 9(4): 680–685.

Zhong, J. & Huang, S. (2006). An enhanced multiplicative spread spectrum watermarking scheme, *IEEE Transactions on Circuits and Systems for Video Technology* 16(12): 1491 –1506.

Zou, J., Tie, X., Ward, R. & Qi, D. (2005). Some novel image scrambling methods based on affine modular matrix transformation, *Journal of Info. and Compt. Sci.* 2(1): 223–227.

Zou, J. & Ward, R. (2003). Introducing two new image scrambling methods, *IEEE Pacific Rim Conf. on Comm., Comp. and signal Proces.* 2: 708 – 711 vol.2.

Zou, J., Ward, R. & Qi, D. (2004). A new digital image scrambling method based on fibonacci numbers, *Proc. of Int. Symp. on Cir. and Sys.* 3: 965–968.

5

Signal and Image Denoising Using Wavelet Transform

Burhan Ergen
Fırat University
Turkey

1. Introduction

The wavelet transform (WT) a powerful tool of signal and image processing that have been successfully used in many scientific fields such as signal processing, image compression, computer graphics, and pattern recognition (Daubechies 1990; Lewis and Knowles 1992; Do and Vetterli 2002; Meyer, Averbuch et al. 2002; Heric and Zazula 2007). On contrary the traditional Fourier Transform, the WT is particularly suitable for the applications of non-stationary signals which may instantaneous vary in time (Daubechies 1990; Mallat and Zhang 1993; Akay and Mello 1998). It is crucial to analyze the time-frequency characteristics of the signals which classified as non-stationary or transient signals in order to understand the exact features of such signals (Rioul and Vetterli 1991; Ergen, Tatar et al. 2010). For this reason, firstly, researchers has concentrated on continuous wavelet transform (CWT) that gives more reliable and detailed time-scale representation rather than the classical short time Fourier transform (STFT) giving a time-frequency representation (Jiang 1998; Qian and Chen 1999).

The CWT technique expands the signal onto basis functions created by expanding, shrinking and shifting a single prototype function, which named as mother wavelet, specially selected for the signal under considerations. This transformation decomposes the signal into different scales with different levels of resolution. Since a scale parameter shrinking or expanding the mother wavelet in CWT, the result of the transform is time-scale representation. The scale parameter is indirectly related to frequency, when considered the center frequency of mother wavelet.

A mother wavelet has satisfy that it has a zero mean value, which require that the transformation kernel of the wavelet transform compactly supports localization in time, thereby offering the potential to capture the spikes occurring instantly in a short period of time (Mallat 1989; Rioul and Vetterli 1991; Akay and Mello 1998).

A wavelet expansion is representation of a signal in terms of an orthogonal collection of real-valued generated by applying suitable transformation to the original selected wavelet. The properties and advantages of a family of wavelets depend upon the mother wavelet features. The expansion is formed by two dimensional expansion of a signal and thus provides a time-frequency localization of the input signal. This implies that most of the energy of the signal will be captured a few coefficient. The basis functions in a wavelet transform are produced from the mother wavelet by scaling and translation operations. When the scaling is chosen as power of two, this kind of wavelet transform is called dyadic-orthonormal wavelet transform, which makes a way for discrete wavelet transform (Zou and Tewfik 1993; Blu 1998) . If the chosen mother wavelet has orthonormal properties, there is no redundancy in the discrete wavelet transforms. In addition, this provides the multiresolution algorithm decomposing a signal into scales with different time and frequency resolution (Mallat 1989; Daubechies 1990).

The fundamental concept involved in mutiresolution is to find average features and details of the signal via scalar products with scaling signals and wavelets. The spikes in signal are typically of high frequency and it is possible discriminate the spikes with other noises through the decomposition of multiresolution into different levels. The differences between mother wavelet functions (e.g. Haar, Daubechies, Symlets, Coiflets, Biorthogonal and etc.) consist in how these scaling signals and the wavelets are defined (Zou and Tewfik 1993; Blu 1998; Ergen, Tatar et al. 2010).

The continuous wavelet transform is computed by changing the scale of the mother wavelet, shifting the scaled wavelet in time, multiplying by the signal, and integrating over all times. When the signal to be analyzed and wavelet function are discredited, the CWT can be realized on computer and the computation time can be significantly reduced if the redundant samples removed respect to sampling theorem. This is not a true discrete wavelet transform. The fundamentals of discrete wavelet transform goes back to sub-band coding theorem (Fischer 1992; Vetterli and Kova evi 1995; Vetterli and Kovacevic 1995). The sub-band coding encodes each part of the signal after separating into different bands of frequencies. Some studies have made use of wavelet transform as a filter bank in order to separate the signal.

After discovering the signal decomposition of a signal into frequency bands using discrete wavelet transform, the DWT has found many application area, from signal analysis to signal compression (Chang and Kuo 1993; Qu, Adam et al. 2003; He and Scordilis 2008).

The one of the first application of the DWT is the denoising process, which aims to remove the small part of the signal assumed as noise (Lang, Guo et al. 1996; Simoncelli and Adelson 1996; Jansen 2001). All kind of the signal obtained from the physical environment has contains more or less disturbing noise. Therefore, wavelet denoising procedure has applied many one or two dimensional signal after particularly soft or hard thresholding methods had proposed (Donoho and Johnstone 1994; Donoho 1995). Such signals some time are one-dimensional simple power or control signals (Sen, Zhengxiang et al. 2002; Giaouris, Finch et al. 2008) as well as more complex medical images (Wink and Roerdink 2004; Pizurica, Wink et al. 2006). Especially, wavelet denoising has found an application field about image processing recently (Nasri and Nezamabadi-pour 2009; Chen, Bui et al. 2010; Jovanov, Pi urica et al. 2010).

2. Noise consideration

A signal or an image is unfortunately corrupted by various factors which effects as noise during acquisition or transmission. These noisy effects decrease the performance of visual and computerized analysis. It is clear that the removing of the noise from the signal facilitate the processing. The denoising process can be described as to remove the noise while retaining and not distorting the quality of processed signal or image (Chen and Bui 2003; Portilla, Strela et al. 2003; Buades, Coll et al. 2006). The traditional way of denoising to remove the noise from a signal or an image is to use a low or band pass filter with cut off frequencies. However the traditional filtering techniques are able to remove a relevant of the noise, they are incapable if the noise in the band of the signal to be analyzed. Therefore, many denoising techniques are proposed to overcome this problem.

The algorithms and processing techniques used for signals can be also used for images because an image can be considered as a two dimensional signal. Therefore, the digital signal processing techniques for a one dimensional signal can be adapted to process two dimensional signals or images.

Because the origin and nonstationarity of the noise infecting in the signal, it is difficult to model it. Nevertheless, if the noise assumed as stationary, an empirically recorded signal that is corrupted by additive noise can be represented as;

$$y(i) = x(i) + \sigma\varepsilon(i) , \; i = 0,1,...,n-1 \tag{1}$$

Where $y(i)$ noisy signal, $x(i)$ is noise free actual signal and $\varepsilon(i)$ are independently normal random variables and σ represents the intensity of the noise in $y(i)$. The noise is usually modeled as stationary independent zero-mean white Gaussian variables (Moulin and Liu 1999; Alfaouri and Daqrouq 2008).

When this model is used, the objective of noise removal is to reconstruct the original signal $x(i)$ from a finite set of $y(i)$ values without assuming a particular structure for the signal. The usual approach to noise removal models noise as high frequency signal added to an original signal. These high frequencies can be bringing out using traditional Fourier transform, ultimately removing them by adequate filtering. This noise removal technique conceptually clear and efficient since depends only calculating DFT (Discrete Fourier Transform)(Wachowiak, Rash et al. 2000).

However, there is some issue that must be under consideration. The most prominent having same frequency as the noise has important information in the original signal. Filtering out these frequency components will cause noticeable loss of information of the desired signal when considered the frequency representation of the original signal. It is clear that a method is required in order to conserve the prominent part of the signal having relatively high frequencies as the noise has. The wavelet based noise removal techniques have provided this conservation of the prominent part.

3. Discrete Wavelet Transform (DWT) and Wavelet Packet Decomposition

The wavelet transform has become an essential tool for many applications. However, the wavelet transform has been presented a method representing a time-frequency method, continuous wavelets transform (CWT), and the wavelet transform generally has used for the decomposition of the signal into high and low frequency components. The wavelet coefficient represents a measure of similarity in the frequency content between a signal and a chosen wavelet function. These coefficients are computed as a convolution of the signal and the scaled wavelet function, which can be interpreted as a dilated band-pass filter because of its band-pass like spectrum (Valens ; Rioul and Vetterli 1991) .

In practice, the wavelet transform is implemented with a perfect reconstruction filter bank using orthogonal wavelet family. The idea is to decompose the signal into sub-signals corresponding to different frequency contents. In the decomposition step, a signal is decomposed on to a set of orthonormal wavelet function that constitutes a wavelet basis (Misiti, Misiti et al.). The most common wavelets providing the ortogonality properties are daubechies, symlets, coiflets and discrete meyer in order to provide reconstruction using the fast algorithms (Beylkin, Coifman et al. 1991; Cohen, Daubechies et al. 1993).

The use of wavelet transform as filter bank called as DWT (Discrete Wavelet Transform). The DWT of a signal produces a non-redundant restoration, which provides better spatial and spectral localization of signal formation, compared with other multi-scale representation such as Gaussian and Laplacian pyramid. The result of the DWT is a multilevel decomposition, in which the signal is decomposed in *'approximation'* and *'detail'* coefficients at each level (Mallat 1989). This is made through a process that is equivalent to low-pass and high passes filtering, respectively.

As stated previous section, the wavelet transform is firstly introduced for the time-frequency analysis of transient continuous signals, and then extended to the theory of multi-resolution wavelet transform using FIR filter approximation. This managed using the dyadic form of CWT. In dyadic form, the scaling function is chosen as power of two. And then, the discrete wavelets $\psi_{m,n}(t) = 2^{-m/2}\psi(2^{-m}t - n)$ used in multi-resolution analysis constituting an orthonormal basis for $L^2(\Re)$ (Vetterli and Herley 1992; Donoho and Johnstone 1994).

If a signal, $x(t)$, decomposed into low and high frequency components, that they are respectively named as approximation coefficients and detail coefficients, $x(t)$ reconstructed as;

$$x(t) = \sum_{m=1}^{L}\left[\sum_{k=-\infty}^{\infty} D_m(k)\psi_{m,k}(t) + \sum_{k=-\infty}^{\infty} A_l(k)\phi_{l,k}(t)\right] \qquad (2)$$

Where $\psi_{m,k}(t)$ is discrete analysis wavelet, and $\phi_{l,k}(t)$ is discrete scaling, $D_m(k)$ is the detailed signal at scale 2^m, and $A_l(k)$ is the approximated signal at scale 2^l. $D_m(k)$ and $A_l(k)$ is obtained using the scaling and wavelet filters (Mallat 1999).

$$h(n) = 2^{-1/2}\langle\phi(t),\phi(2t - n)\rangle$$
$$g(n) = 2^{-1/2}\langle\psi(t),\phi(2t - n)\rangle \qquad (3)$$
$$= (-1)^n h(1 - n)$$

The wavelet coefficient can be computed by means of a pyramid transfer algorithm. The algorithms refer to a FIR filter bank with low-pass filter **h**, high-pass filter **g**, and down sampling by a factor 2 at each stage of the filter bank. Fig. 1 shows the tree structure of DWT decomposition for three levels. DWT decomposition leads to a tree structure as shown in Fig. 1, where approximation and detail coefficients are presented.

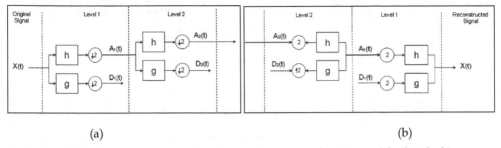

(a)   (b)

Fig. 1. The DWT decomposition and reconstruction steps of a 1D signal for level of 2;
a. Decomposition, b. Reconstruction

In this figure, $\downarrow 2$ and $\uparrow 2$ refers to down sampling and up sampling, respectively. This decomposition sometimes called as sub-band coding. The low pass filter produces the approximation of the signal, and the high pass filers represent the details or its high frequency components. The decomposition successively can be applied on the low frequency components, approximation coefficients, in DWT.

Whereas the successive decomposition is applied on the approximation coefficients only as in the DWT, the decomposition may be applied on both sub part of the signal, approximation coefficients and detail coefficients. If the decomposition is applied on the both sides, approximation and details, this kind of decomposition called as wavelet packet transform or wavelet packet tree decomposition. Fig. 2 represents wavelet packet decomposition and reconstruction.

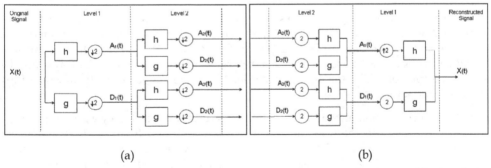

(a)   (b)

Fig. 2. The wavelet packet decomposition and reconstruction steps of a 1D signal for level of 2;
a. Decomposition, b. Reconstruction

In 2D case, the image signal is considered as rows and columns as if they are one dimentional signals. In DWT, firstly the each rows of the image is filtered, then the each columns are filtered as in 1D case. Figure 3 demonstare the decompositon of an image for one level. As in signal decomposition, after each filtering, the subsampling is realized. The result of this process gives four images; approximation, horizantal details, vertical details and diagonal details. Because of subsampling after each filtering, the result subimages of the original image has the quarter size of the original image.

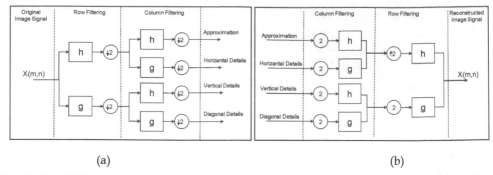

(a)   (b)

Fig. 3. The DWT decomposition and reconstruction steps of a 2D image signal for level of 2; a.Decomposition, b. Reconstruction

4. Thresholding and threshold estimation techniques

The simpler way to remove noise or to reconstruct the original signal from a contaminated signal, in case of 1D or 2D, using the wavelet coefficients which are the result of decomposition in wavelet transform, is to eliminate the small coefficient associated to the noise. After updating the coefficients by removing the small coefficients assuming as noise, the original signal can be obtained by the reconstruction algorithm using the noise free coefficients. Because it is usually considered that the noise has high frequency coefficients, the elimination of the small coefficient generally applied on the detail coefficients after the decomposition. Indeed, the main idea of the wavelet denoising to obtain the ideal components of the signal from the noisy signal requires the estimation of the noise level. The estimated noise level is used in order to threshold the small coefficient assumed as noise.

The procedure of the signal denoising based on DWT is consist of three steps; decomposition of the signal, thresholding and reconstruction of the signal. Several methods use this idea proposed and implements it in different ways. When attempting to decrease the influence of noise wavelets coefficient, it is possible to do this in particular ways, also the need of information of the underlying signal leads to different statistical treatments of the available information.

In the linear penalization method every wavelet coefficient is affected by a linear shrinkage particular associated to the resolution level of the coefficient. It can be said that linear thresholding is appropriate only for homogeny signals with important levels of regularity. The wavelet thresholding or shrinkage methods are usually more suitable. Since the work of Donoho and Johnstone (Donoho and Johnstone 1994), there has been a lot of research on the way of defining the threshold levels and their type. Donoho and Johnstone proposed a nonlinear strategy for thresholding. In their approaches, the thresholding can be applied by implementing either hard or soft thresholding method, which also called as shrinkage.

In the hard thresholding, the wavelet coefficient below a give value are stetted to zero, while in soft thresholding the wavelet coefficient are reduced be a quantity to the thresh value. The threshold value is the estimation of the noise level, which is generally calculated from the standard deviation of the detail coefficient (Donoho 1995). Fig. 4 indicates the two types of thresholding, which can be expressed analytically as;

$$Hard\ threshold: \begin{cases} y = x & if\ |x| > \lambda \\ y = 0 & if\ |x| < \lambda \end{cases} \tag{4}$$

$$Soft\ threshold: \{y = sign(x)(|x| - \lambda) \tag{5}$$

Where x is the input signal, y is the signal after threshold and λ is the threshold value, which is critical as the estimator leading to destruction, reduction, or increase in the value of a wavelet coefficient.

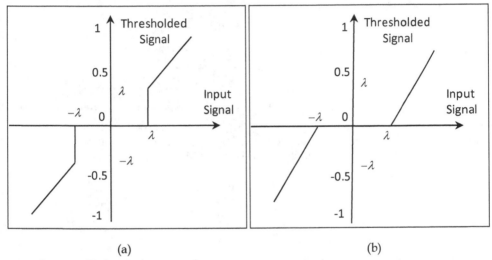

(a)   (b)

Fig. 4. Threshold types; a. Hard, b. Soft.

Hard thresholding method does not affect on the detail coefficients that grater the threshold level, whereas the soft thresholding method to these coefficients. There are several considerations about the properties and limitation of these two strategies. However the hard thresholding may be unstable and sensitive even to small changes in the signal, the soft thresholding can create unnecessary bias when the true coefficients are large. Although more sophisticated methods has been proposed to overcome the drawbacks of the described nonlinear methods, it is still the most efficient and reliable methods are still the hard and soft thresholding techniques (Donoho 1995).

One important point in thresholding methods is to find the appropriate value for the threshold. Actually, many approaches have been proposed for calculating the threshold value. But, all the approaches require the estimation of noise level. However the standard deviation of the data values may be use as an estimator, Donoho proposed a good estimator σ for the wavelet denoising given as;

$$\sigma = \frac{median(d_{L-1,k})}{0.6745}, \quad k = 0,1,...,2^{L-1} - 1 \tag{6}$$

where L denotes the number of decomposition levels. As mentioned above, this median selection made on the detail coefficient of the analyzed signal.

The most known threshold selection algorithms are minimax, universal and rigorous sure threshold estimation techniques (Donoho and Johnstone 1994; Donoho and Johnstone 1998).

The *minimax* threshold value λ_M proposed by Donoho consists an optimal threshold that derived from minimizing the constant term in an upper bound of the risk involved in the estimation. The proposed threshold depends of the available data and also takes into account the noise level contaminating the signal. The optimal threshold is defined as;

$$\lambda_M = \sigma\lambda_n^*$$ (7)

where λ_n^* is defined as the value of λ and satisfying as;

$$\lambda_n^* = \inf_\lambda \sup_d \left\{ \frac{R_\lambda(d)}{n^{-1} + R_{oracle}(d)} \right\}$$ (8)

where $R_\lambda(d) = E(\delta_\lambda(d) - d)^2$ and $R_{oracle}(d)$ named as oracle used to account for the risk associated to the modification of the value of a given wavelet coefficient. Two oracles are considered, the diagonal liner projection (DLP) and the diagonal linear shrinker (DLS)(Donoho and Johnstone 1994). The ideal risks for these oracles are given by

$$R_{oracle}^{DLP}(d) = \min(d^2, 1)$$ (9)

$$R_{oracle}^{DLS}(d) = \frac{d^2}{d^2 + 1}$$ (10)

The minimax method is used in statistics to design estimator. The minimax estimator is realizes the minimum of the maximum mean square error, over a given set of functions. Another proposed threshold estimator by Donoho is the *universal* threshold, or global threshold, as an alternative to the minimax threshold, however it uses a fixed threshold form given as;

$$\lambda_U = \sigma\sqrt{2\log(n)}$$ (11)

Where n denotes the length of the analyzed signal and σ is given by Eq. (6). The advantage of this thresholding appears in software implementation due to easy to remember and coding. Additionally, this threshold estimator ensures that every sample in the wavelet transform in which the underlying function is exactly zero will be estimated as zero.

Again another common estimator is Rigorous Sure (rigresure) threshold proposed by Donoho. This threshold describes a scheme which uses a threshold λ at each resolution level l of the wavelet coeffient. The Rigorous Sure, also known as *SureShrink*, uses the Stein's Unbiased Risk Estimate criterion to obtain unbiased estimate. The threshold is given as follows;

$$\lambda_S = \arg\min_{0<\lambda<\lambda_U} Sure\left(\lambda, \frac{S(a,b)}{\sigma} \right)$$ (12)

Where Sure is defined as

$$Sure(\lambda, X) = n - 2 \cdot \Theta\{i : |X_i \le \lambda|\} + \left[\min(|X_i|, \lambda\right]^2 \qquad (13)$$

Where the operator $\Theta(\cdot)$ returns the cardinality of the set $\{i : |X_i \le \lambda|\}$, it is found that Sure is an un biased estimate of the $l^2 - risk$.

5. Denoising application examples

5.1 Comparison assessments

The best way to test the effect of noise on a signal is to add a Gaussian white noise, in which case its values independently and identically distributed (i.i.d) Gaussian real values. After the denoising process, the performance can be measure by comparing the denoised signal and the original signal. However, many methods have been proposed to measure the performance of denoising algorithms, the signal to noise ratio (SNR) and peak signal to noise ratio (PSNR) has generally accepted to measure the quality of signal and images, respectively. For one dimensional signal, measuring the performance of the denoising method by calculation of the residual SNR given as;

$$SNR = 10\log_{10}\left(\sum_{n=0}^{N-1} x^2(n) \middle/ \sum_{n=0}^{N-1} \left(\overline{x}(n) - x^r(n]\right)\right)^2\right) \qquad (14)$$

where $x(n)$ is the original signal, $x_r(n)$ is the denoised signal and $\overline{x}(n)$ refers to the mean value of $x(n)$.

In order to measure the quality of image, it is generally used PSNR, which given as;

$$PSNR = 10\log_{10}\left(L\middle/ \sum_{n=0}^{N-1}\sum_{m=0}^{M-1}\left(\overline{x}(n,m) - x^r(n,m)\right)^2\right) \qquad (15)$$

where L denotes the quantized gray level of the images, $x(n)$ is original images, $\overline{x}(n,m)$ is the mean value of $x(n)$, and $x^r(n,m)$ refers to reconstructed image. In order to get visible alteration on signal, the power of noise should be chosen adequately. Indeed, SNR is usually the most important measure rather than the power of noise, when taking into consideration that the power of the signal to denoise can be varied. When the SNR is chosen above 3dB, it is generally enough to get the visible corruption.

5.2 Phonocardiogram denoising

The records of the acoustical vibrations produced by heart, acquired through microphones from human chest, called phonocardiogram (PCG), consist of the heart sounds and the murmurs. This records of acoustic signals are unfortunately disturbed by various factors which effecting as noise. These effects decrease the performance of visual and computerized analysis (Akay, Semmlow et al. 1990; Ergen, Tatar et al. 2010).

The respiration sounds by lung mechanical actions, patient movement, and improper contacts of microphone to the skin, and external noises from the environments are added as noise signal into PCG records. The traditional method to remove the noise from a PCG signal is to use a low or band pass filter with cut off frequencies. However the filtering techniques are able to remove a relevant of the noise, they are incapable if the noise in the band of the signal to be analyzed.

The frequency components of a normal PCG signals can be rise up 200Hz, and the energy of the most significant components concentrates around the frequency band 100-150Hz (Ergen, Tatar et al. 2010). The frequency bands of the signal are very important when we use the denoising technique using DWT approaches. Because the DWT approaches decomposes the signal into frequency bands to eliminate the small detail components assumed as noise, the decomposition level reflects directly on the frequency components that cause the smoothed version of the signal.

As stated previous section, the most reasonable way to determine the effectiveness of denoising method is to compare an original signal and the denoised signal obtained from its noise added form. Therefore, here, we will use the noise added signal to examine the effectiveness of wavelet denoising method through the comparison between the original signal and the denoised signal (or reconstructed) signal. Figure 5.a shows a PCG during cardiac period and its noise added form.

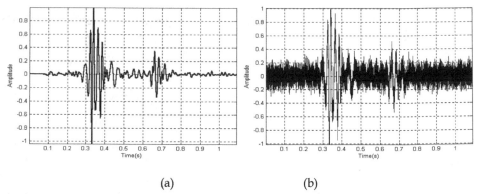

(a)   (b)

Fig. 5. Wavelet denoising of a PCG signal, a) Original signal, b) Noisy signal

The result of the DWT is a multilevel decomposition, in which the signal is decomposed in 'approximation' and 'detail' coefficients at each level. This is made through a process that is equivalent to low-pass and high passes filtering, respectively. DWT decomposition leads to a tree structure as shown in Fig. 6, where approximation and detail coefficients are presented.

The approximation coefficients and the detail coefficients of the noisy signal for the decomposition level of one and two are given Figure 7. In Fig. 7c and Fig. 7d are the results of the decomposition of the approximation coefficient at level one, which represented in Fig. 8a.

As an example of denoising process for PCG signal, the denoised signal and the difference between the original signal and the denoised signal are given in Fig. 7, respectively. *'symlet8'*, *'rigresure'* and *'soft thresholding'* parameters are used in the denoising process. When we compared even the original the original signal and the denoised signal visually, the wavelet denoising process has a quite success.

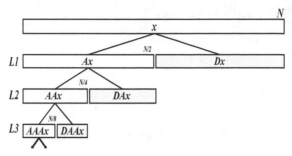

Fig. 6. The approximation and the detailed coefficients in the tree structure of the DWT.

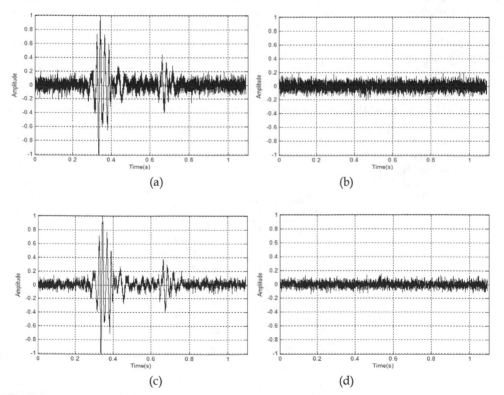

Fig. 7. Decomposition of the noisy signal,
a) Approximation coefficients at level one, b) Detail coefficients at level one,
c) Approximation coefficients at level two, b) Detail coefficients at level two.

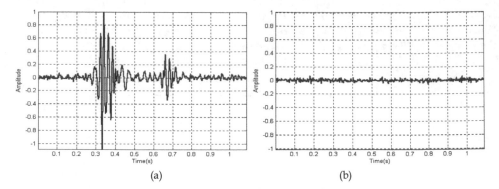

(a)   (b)

Fig. 8. Denoised signal, a) Denoised signal, d) Difference between the original and the denoised signal.

Also, the effected components in DWT decomposition are related to not only decomposition level but also sampling frequency. The decomposition level influences the frequency bands by dividing the sampling frequency respect the power of two. When we choose the decomposition level is as five, the interested frequencies are about 300Hz while the sampling frequency is 11.5KHz.

Therefore, the most important factor determining the SNR level is the level of the decomposition. Table 1 presents the SNR results respect to the decomposition level by using *symlet8* and *rigresure* estimation for hard and soft thresholding as denoising parameter. For the both tresholding techniques, it is seen that the highest SNR value obtained when the decomposition level is five. If the decomposition level is chosen too high, the thresholding will effect on the main frequencies of the original signal. Thus, the SNR has lower values for the level higher than five.

Level	Hard	Soft
1	8.1209	7.8843
2	11.1471	10.9218
3	14.3251	14.0031
4	17.2973	16.9275
5	20.1305	19.4396
6	13.2248	13.2472
7	12.1531	9.8726
8	10.8010	8.3255
9	10.4986	8.1632
10	10.4912	8.1593

Table 1. SNR level respect to the depth of decomposition.

The other parameters to obtain best SNR level are the kind of the wavelet and the thresholding rule. Table 2 presents the SNR levels using different wavelet when the decomposition level is five. In table 2, there is no significant difference in SNR in terms of wavelet types.

Wavelet Type	Hard	Soft
Daubechies2	16.5378	16.5057
Daubechies3	18.9391	18.8353
Daubechies4	19.8138	19.8002
Daubechies5	19.8747	19.7425
Symlet2	16.3487	16.4181
Symlet3	18.5401	18.7874
Symlet4	19.5732	19.8002
Symlet5	19.4795	19.5458
Coiflet1	16.7746	16.7658
Coiflet2	19.4866	19.4501
Coiflet 3	19.7812	19.6252
Discrete Meyer	19.9018	19.7154

Table 2. SNR values respect to wavelet types (Rigrsure, level=5)

Nevertheless, it is attracting that the mother wavelets having high oscillation number produces better SNR results. For instance, the symlet wavelet having eight oscillations in its mother wavelet produces better SNR level than the lower ones. In this case, it can be say that the choice of the very lower oscillation frequency to avoid the computational complexity of the wavelet causes the lower SNR results.

When the performance of the noise estimation techniques is considered in the respect of the decomposition level and the initial SNR level, the estimation techniques show the same performance for the level five respects to the initial SNR level. For the comparison, the initial SNR level before denoising is increased from 1dB to 30dB, and the result SNR level after denoising is calculated. Fig. 9 presents a comparison of the four noise estimation methods for level five and level six when 'symlet8' used.

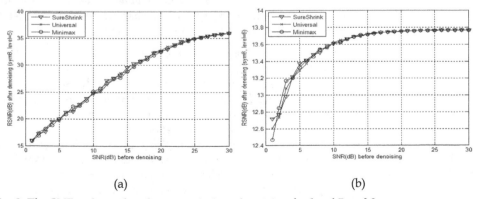

(a)   (b)

Fig. 9. The SNR values after denoising before denoising for level 5 and 8.

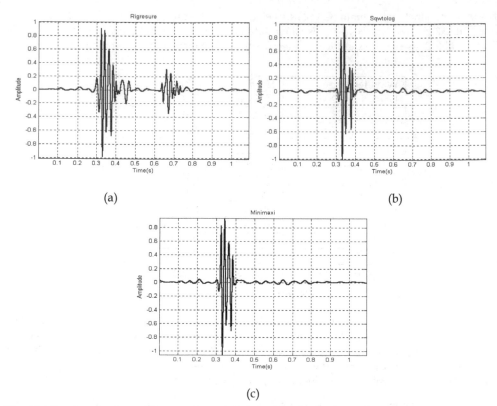

(a)   (b)

(c)

Fig. 10. The denoised signal using three different threshold rules at level eight.

We have observed no distinguishing evidence among the noise level estimation methods until level six. After this level, rigresure method has produced better SNR values. And it is observed that rigresure preserve the second heart sound in PCG signals while the other methods destroying. This situation is clearly seen in Fig. 10. The signal part related to second heart sound taking place at around 0.7s in Fig.10a is not able to seen in Fig. 10b and Fig. 10c. This shows that the rigresure preserve the main characteristic of the signal. Therefore, we can conclude that the rigresure is the better noise estimation method.

A level-dependent scaling of the thresholds was used to remove Gaussian white noise from the signal. Although it could not found evidence that a single wavelet was the best suited for denoising PCG signal, some wavelets used in this study were slightly better than the others. We conclude that reasonable decomposition level is absolutely depending on the sampling frequency and the frequency band of the signal. Just in this study, the decomposition level of 5 produced reasonable results because the frequency band of a normal PCG signal is around 150-200Hz and the sampling frequency is 11.5KHz. Since the noise level method is one of the important parameter in wavelet denoising, it is examined for different levels. We have not seen any noteworthy differences in the methods from level 1 to level 6. After this level, rigresure method has showed superiority to the other methods in terms of SNR level. Consequently, it is determined that the wavelet type is not very important if the oscillation number is not very low, the decomposition level is absolutely depends on the frequency band of the PCG signal and its sampling frequency, and rigresure method is best of the noise estimation techniques.

5.3 Image denoising

All digital images contain some degree of noise due to the corruption in its acquisition and transmission by various effects. Particularly, medical image are likely disturbed by a complex type of addition noise depending on the devices which are used to capture or store them. No medical imaging devices are noise free. The most commonly used medical images are received from MRI (Magnetic Resonance Imaging) and CT (Computed Tomography) equipments. Usually, the addition noise into medical image reduces the visual quality that complicates diagnosis and treatment.

Because the wavelet transform has an ability to capture the energy of a signal in few energy transform values, the wavelet denoising technique is very effective as stated previous parts. As stated previous sections, when an image is decomposed using wavelet transform, the four sub-images are produced, approximation, horizontal details, vertical details and diagonal details. Fig. 11 represents a sample medical image which belongs to a patient having cranial trauma and its four subimages when decomposed for one level using DWT. This image has acquired from a BT device. A noise added MRI image and its denoised form using wavelet denoising procedure is given Fig. 12. The added noise has Gaussian distribution, and symlet6, decomposition level of two, hard thresholding are chosen as wavelet denoising parameters.

Fig. 11. Decomposition of a sample medical image; original, approximation, horizontal details, vertical details, and diagonal details in left to right.

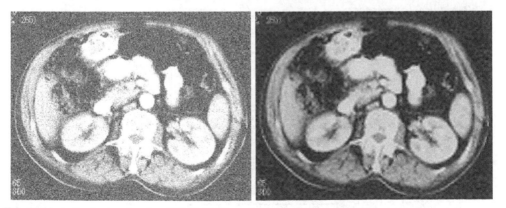

Fig. 12. A noisy image having PSNR 62dB and its denoised version.

Quantitatively assessing the performance in practical image application is complicated issue because the ideal image is normally unknown. Therefore the rational approach is to use known images for the tests, as in other image processing applications, in order to test the performance of the wavelet denoising methods like one dimensional signal denoising. Figure 13 represents the medical test images to be used.

Here, we use again a classical comparison receipt based on noise simulation. The comparison can be realized on the result reconstructed image and the original image after adding Gaussian white noise with known power to the original signal. Then it will be computed the best image recovered from the noisy one for each method. Firstly, we should determine the effective decomposition level because the most important factor in wavelet denoising is decomposition level. For this purpose, a noise added image will be used to obtain how the performance is changing respect to the decomposition level. The recovering process is made on the test image given in Fig 11, on which a Gaussian noise added to be PSNR is 62dB. The noisy image and a sample recovered or denoised is given Fig. 12a and Fig. 12b, respectively. The PSNR values after denosing process is given Table 3. In this denoising process, the symlet6 and universal thresholding is chosen as mother wavelet and noise level estimator.

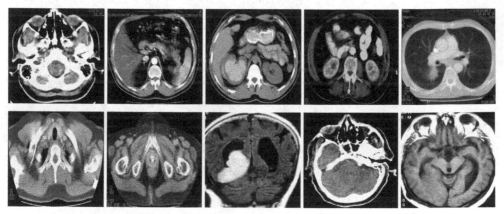

Fig. 12. Medical test images.

Level	PSNR
1	68.1196
2	69.3269
3	70.5006
4	70.7768
5	68.6232
6	68.8183
7	68.7272
8	69.8037
9	66.8912
10	66.3877

Table 3. PSNR values respect to decomposition level after DWT denoising.

The best PSNR is obtained at the decomposition level of two. As can be seen in Table 3, the result PSNR value is decreasing if the decomposition level getting higher. The wavelet transform concern the main component of the original signal when the decomposition level is increased. If the higher decomposition level is used, the thresholding can eliminate some coefficients of the original signal, as in 1D signal denoising process. Therefore, to increase the decomposition level too high will decrease the PSNR after an optimal level and also increase the complexity of decomposition. In further part of the study, the decomposition level is chosen as two because the performance of the DWT denoising obtained at this level. Another question about the performance of the wavelet denoising is if it is dependent on the content or the distribution of the coefficient of the image. We can answer the question by applying the denoising algorithm on different images. Table 4 represents the PSNR values respect to the number of the test images given in Fig. 9 after the denoising process.

Number	Noisy Image	Denoised Image (level1)	Denoised Image (level2)	Denoised Image (level3)	Denoised Image (level4)
1	62.0974	68.1252	73.2903	72.9250	70.3792
2	62.1251	67.3979	69.3305	68.4441	67.3593
3	62.1140	67.9648	71.7193	70.4829	68.8435
4	62.0942	67.9819	72.2531	72.4830	70.8092
5	62.0974	67.0273	69.5873	69.9803	69.2444
6	62.1023	67.8774	71.7282	71.4382	70.0891
7	62.1138	67.6268	70.6594	70.8403	69.8362
8	62.0995	68.1391	73.7535	74.2233	71.6437
9	62.1224	67.9712	71.3191	69.4574	67.9060
10	62.1070	67.9048	71.0798	69.1241	67.4048
Mean	62.1069	67.8016	71.4721	70.9399	69.3515
Standard Deviation	0,176	0.3521	1.4230	1.8322	1.4656

Table 4. PSNR's respect to image number, mean and standart deviation.

6. Conclusion

The wavelet denoising techniques offers high quality and flexibility for the noise problem of signals and image. The performances of denoising methods for several variations including thresholding rules and the type of wavelet were examined in the examples in order to put forward the suitable denoising results of the methods. The comparisons have made for the three threshold estimation methods, wavelet types and the threshold types. The examinations have showed that most important factor in wavelet denoising is what the decomposition level is rather than the wavelet type, threshold type or the estimation of threshold value.

However, someone has not seen any noteworthy differences in the methods from level one to level six, after this level, rigresure method has showed a better performance than the other methods in terms of SNR level. Consequently, it is determined that the wavelet type is not very important if the oscillation number is not very low, the decomposition level is absolutely depends on the frequency band of the signal to be analyzed and its sampling frequency.

7. References

Akay, M. and C. Mello (1998). Wavelets for biomedical signal processing, IEEE.

Akay, M., J. Semmlow, et al. (1990). "Detection of coronary occlusions using autoregressive modeling of diastolic heart sounds." Biomedical Engineering, IEEE Transactions on 37(4): 366-373.

Alfaouri, M. and K. Daqrouq (2008). "ECG signal denoising by wavelet transform thresholding." American Journal of Applied Sciences 5(3): 276-281.

Beylkin, G., R. Coifman, et al. (1991). "Fast wavelet transforms and numerical algorithms I." Communications on pure and applied mathematics 44(2): 141-183.

Blu, T. (1998). "A new design algorithm for two-band orthonormal rational filter banks and orthonormal rational wavelets." Signal Processing, IEEE Transactions on 46(6): 1494-1504.

Buades, A., B. Coll, et al. (2006). "A review of image denoising algorithms, with a new one." Multiscale Modeling and Simulation 4(2): 490-530.

Chang, T. and C. C. J. Kuo (1993). "Texture analysis and classification with tree-structured wavelet transform." Image Processing, IEEE Transactions on 2(4): 429-441.

Chen, G. and T. Bui (2003). "Multiwavelets denoising using neighboring coefficients." Signal Processing Letters, IEEE 10(7): 211-214.

Chen, G., T. Bui, et al. (2010). "Denoising of three dimensional data cube using bivariate wavelet shrinking." Image Analysis and Recognition: 45-51.

Cohen, A., I. Daubechies, et al. (1993). "Wavelets on the interval and fast wavelet transforms." Applied and Computational Harmonic Analysis 1(1): 54-81.

Daubechies, I. (1990). "The wavelet transform, time-frequency localization and signal analysis." Information Theory, IEEE Transactions on 36(5): 961-1005.

Do, M. and M. Vetterli (2002). Texture similarity measurement using Kullback-Leibler distance on wavelet subbands, IEEE.

Donoho, D. L. (1995). "Denoising by soft-thresholding." IEEE Trans. Inform. Theory 41(3): 613-627.

Donoho, D. L. and I. M. Johnstone (1994). "Ideal spatial adaptation via wavelet shrinkage." Biometrika 81(3): 425-455.

Donoho, D. L. and I. M. Johnstone (1998). "Minimax estimation via wavelet shrinkage." Annals of statistics: 879-921.

Donoho, D. L. and J. M. Johnstone (1994). "Ideal spatial adaptation by wavelet shrinkage." Biometrika 81(3): 425.

Ergen, B., Y. Tatar, et al. (2010). "Time-frequency analysis of phonocardiogram signals using wavelet transform: a comparative study." Computer Methods in Biomechanics and Biomedical Engineering 99999(1): 1-1.

Fischer, T. R. (1992). "On the rate-distortion efficiency of subband coding." Information Theory, IEEE Transactions on 38(2): 426-428.

Giaouris, D., J. W. Finch, et al. (2008). "Wavelet denoising for electric drives." Industrial Electronics, IEEE Transactions on 55(2): 543-550.

He, X. and M. S. Scordilis (2008). "Psychoacoustic music analysis based on the discrete wavelet packet transform." Research Letters in Signal Processing 2008: 1-5.

Heric, D. and D. Zazula (2007). "Combined edge detection using wavelet transform and signal registration." Image and Vision Computing 25(5): 652-662.

Jansen, M. (2001). Noise reduction by wavelet thresholding, Springer USA.

Jiang, Q. (1998). "Orthogonal multiwavelets with optimum time-frequency resolution." Signal Processing, IEEE Transactions on 46(4): 830-844.

Jovanov, L., A. Pi urica, et al. (2010). "Fuzzy logic-based approach to wavelet denoising of 3D images produced by time-of-flight cameras." Optics Express 18(22): 22651-22676.

Lang, M., H. Guo, et al. (1996). "Noise reduction using an undecimated discrete wavelet transform." Signal Processing Letters, IEEE 3(1): 10-12.

Lewis, A. S. and G. Knowles (1992). "Image compression using the 2-D wavelet transform." Image Processing, IEEE Transactions on 1(2): 244-250.

Mallat, S. and Z. Zhang (1993). "Matching pursuits with time-frequency dictionaries." IEEE Transactions on signal processing 41(12): 3397-3415.

Mallat, S. G. (1989). "A theory for multiresolution signal decomposition: The wavelet representation." Pattern Analysis and Machine Intelligence, IEEE Transactions on 11(7): 674-693.

Mallat, S. G. (1999). A wavelet tour of signal processing, Academic Pr.

Meyer, F., A. Averbuch, et al. (2002). "Fast adaptive wavelet packet image compression." Image Processing, IEEE Transactions on 9(5): 792-800.

Misiti, M., Y. Misiti, et al. "Wavelet Toolbox(tm) 4." Matlab User's Guide, Mathworks.

Moulin, P. and J. Liu (1999). "Analysis of multiresolution image denoising schemes using generalized Gaussian and complexity priors." Information Theory, IEEE Transactions on 45(3): 909-919.

Nasri, M. and H. Nezamabadi-pour (2009). "Image denoising in the wavelet domain using a new adaptive thresholding function." Neurocomputing 72(4-6): 1012-1025.

Pizurica, A., A. M. Wink, et al. (2006). "A review of wavelet denoising in MRI and ultrasound brain imaging." Current Medical Imaging Reviews 2(2): 247-260.

Portilla, J., V. Strela, et al. (2003). "Image denoising using scale mixtures of Gaussians in the wavelet domain." Image Processing, IEEE Transactions on 12(11): 1338-1351.

Qian, S. and D. Chen (1999). "Joint time-frequency analysis." Signal Processing Magazine, IEEE 16(2): 52-67.

Qu, Y., B. Adam, et al. (2003). "Data reduction using a discrete wavelet transform in discriminant analysis of very high dimensionality data." Biometrics 59(1): 143-151.

Rioul, O. and M. Vetterli (1991). "Wavelets and signal processing." Signal Processing Magazine, IEEE 8(4): 14-38.

Sen, O., S. Zhengxiang, et al. (2002). "Application of wavelet soft-threshold de-noising technique to power quality detection [J]." Automation of Electric Power Systems 19.

Simoncelli, E. P. and E. H. Adelson (1996). Noise removal via Bayesian wavelet coring, IEEE.

Valens, C. "A Really Friendly Guide to Wavelets. 1999." URL: http://perso. orange. fr /polyvalens/ clemens/ wavelets/ wavelets. html [Last accessed: 13 December 2007].

Vetterli, M. and C. Herley (1992). "Wavelets and filter banks: Theory and design." Signal Processing, IEEE Transactions on 40(9): 2207-2232.

Vetterli, M. and J. Kova evi (1995). Wavelets and subband coding, Citeseer.

Vetterli, M. and J. Kovacevic (1995). Wavelets and Subband Coding. Englewood Clis, NJ: Prentice-Hall.

Wachowiak, M. P., G. S. Rash, et al. (2000). "Wavelet-based noise removal for biomechanical signals: A comparative study." Biomedical Engineering, IEEE Transactions on 47(3): 360-368.

Wink, A. M. and J. B. T. M. Roerdink (2004). "Denoising functional MR images: a comparison of wavelet denoising and Gaussian smoothing." Medical Imaging, IEEE Transactions on 23(3): 374-387.

Zou, H. and A. H. Tewfik (1993). "Parametrization of compactly supported orthonormal wavelets." Signal Processing, IEEE Transactions on 41(3): 1428-1431.

6

A DFT-DWT Domain Invisible Blind Watermarking Techniques for Copyright Protection of Digital Images

Munesh Chandra

DIT School of Engineering, Greater Noida
India

1. Introduction

The great success of Internet and digital multimedia technology have made the fast communication of digital data, easy editing in any part of the digital content, capability to copy a digital content without any loss in quality of the content and many other advantages.

The great explosion in this technology has also brought some problems beside its advantages. The great facility in copying a digital content rapidly, perfectly and without limitations on the number of copies has resulted the problem of copyright protection. Digital watermarking is proposed as a solution to prove the ownership of digital data. A watermark, a secret imperceptible signal, is embedded into the original data in such a way that it remains present as long as the perceptible quality of the content is at an acceptable level. The owner of the original data proves his/her ownership by extracting the watermark from the watermarked content in case of multiple ownership claims

In general, any watermarking scheme (algorithm) consists of three parts.

- The watermark.
- The encoder (insertion algorithm).
- The decoder and comparator (verification or extraction or detection algorithm).

Each owner has a unique watermark or an owner can also put different watermarks in different objects the marking algorithm incorporates the watermark into the object. The verification algorithm authenticates the object determining both the owner and the integrity of the object [1].

1.1 Embedding process

Let us denote an image by I, a signature by $S = s_1, s_2, \ldots$ and the watermarked image by I'. E is an encoder function, it takes an image I and a signature S, and it generates new image which is called watermarked image I', mathematically,

$$E\,(I,\,S) = I' \tag{1}$$

It should be noted that the signature S may be dependent on image I. In such cases, the encoding process described by (1) still holds. The figure 1 illustrates the encoding process [1].

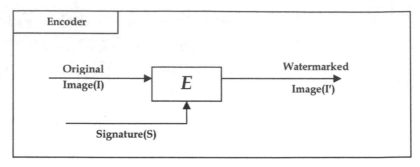

Fig. 1. Encoder

1.2 Extraction process

A decoder function D takes an image J (J can be a watermarked or un-watermarked. image, and possibly corrupted) whose ownership is to be determined and recovers a signature S' from the image [1].

In this process an additional image I can also be included which is often the original and un-watermarked version of J. This is due to the fact that some encoding schemes may make use of the original images in the watermarking process to provide extra robustness against intentional and unintentional corruption of pixels. Mathematically,

$$D (J, I) = S'$$ (2)

In proposed algorithm, original image is not used while extracting watermark from watermarked image and we provide robustness by using some keys.

The extracted signature S' will then be compared with the owner signature sequence by a comparator function C_δ and a binary output decision generated. It is 1 if there is match and 0 otherwise, which can be represented as follows.

$$C_3(S', S) = \begin{cases} 1. & c \le \delta \\ 0. & \text{Otherwise} \end{cases}$$ (3)

Where C is the correlator, $x = c_3(S,S')$. C is the correlation of two signatures and δ is certain threshold. Without loss of generality, watermarking scheme can be treated as a three-tupple (E, D, C_δ). Following figure 2 & figure 3 demonstrate the decoder and the comparator respectively.

A watermark must be detectable or extractable to be useful. Depending on the way the watermark is inserted and depending on the nature of the watermarking algorithm, the method used can involve very distinct approaches. In some watermarking schemes, a watermark can be extracted in its exact form, a procedure we call watermark extraction. In

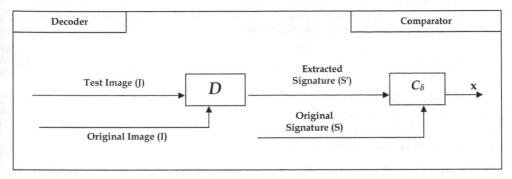

Fig. 2. Decoder

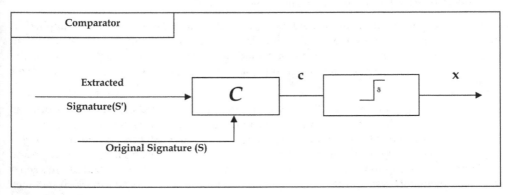

Fig. 3. Comparator

other cases, we can detect only whether a specific given watermarking signal is present in an image, a procedure we call watermark detection. It should be noted that watermark extraction can prove ownership whereas watermark detection can only verify ownership.

The proposed technique extract watermark to prove ownership.

The quality of extracted watermark can also be measured by: *PSNR* (Peak Signal-to-Noise Ratio) and *AR* (Accuracy rate)

PSNR is provided only to give us a rough approximation of the quality of the watermark.

$$PSNR = 10\log_{10}\left(\frac{255^Z}{MSE}\right)dB \qquad (4)$$

Where *MSE* is mean square error of an image with H × W pixels is defined as:

$$MSE = \frac{1}{HXW}\sum_{i=1}^{H}\sum_{j=1}^{W}\left(a_{ij} - \overline{a}_{ij}\right)^2 \qquad (5)$$

Where a_{ij} is the original pixel value and $\overline{a}_{ij}$ is the processed pixel value.

Besides, we utilized the accuracy rate AR to evaluate the robustness of a copyright protection scheme for a specific attack. The formula for AR is shown below:

$$AR = \frac{CP}{NP} \qquad (6)$$

Where NP is the number of pixels of the watermark image and CP is the number of correct pixels in the extracted watermark image.

1.3 Classification of watermarking techniques

Watermarking techniques can be divided into four categories according to the type of document to be watermarked as follows [1]: Text Watermarking, Image Watermarking, Audio Watermarking and Video Watermarking.

In the case of images from implementation point of view, watermarks can be applied in spatial domain and in frequency domain. In Spatial domain, pixels of one and two randomly selected subsets of an image are modified based on perceptual analysis of the original image. In Frequency domain, values of certain frequencies are altered from their original.

According to human perception, digital watermarks can be divided into three categories as follows [2]: Visible, Invisible-robust and Invisible-Fragile. Visible watermark is where the secondary translucent overlaid into the primary content and appears visible on a careful inspection. Invisible-Robust watermark is embedded is such a way changes made to the pixel value are perceptually unnoticed. Invisible –Fragile watermark is embedded in such a way that any manipulation of the content would alter or destroy the watermark. Sometimes another watermarking called dual watermarking is used. Dual watermark is a combination of a visible and an invisible watermark [1]. In this type of watermark an invisible watermark is used as a back up for the visible watermark as clear from the following diagram.

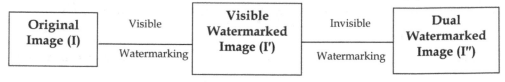

Fig. 4. Schematic representation of dual watermarking

From application point of view, digital watermarking could also be [2]: source based and destination based. In source based a unique watermark identifying the owner is introduced to all the copies of particular content being distributed. Destination based is where each distributed copy gets a unique watermark identifying the particular buyer. Different types of watermarks are shown in the figure. 5.

Current digital image watermarking techniques can be grouped into two major classes: spatial-domain and frequency-domain watermarking techniques [3]. Compared to spatial domain techniques [4], frequency-domain watermarking techniques proved to be more effective with respect to achieving the imperceptibility and robustness requirements of digital watermarking algorithms [5].

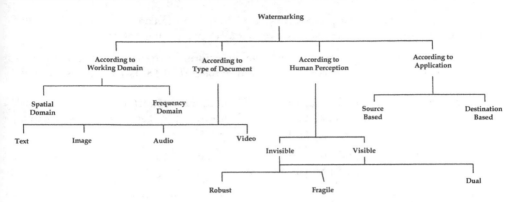

Fig. 5. Types of watermarking techniques

Commonly used frequency-domain transforms include the Discrete Wavelet Transform (DWT), the Discrete Cosine Transform (DCT) and Discrete Fourier Transform (DFT). The host signal is transformed into a different domain and the watermark is embedded in selective coefficients. Here we have described DFT and DWT domain techniques.

1.3.1 Discrete Fourier transform

The Discrete Fourier Transformation (DFT) controls the frequency of the host signal. Energy of watermarking message can be distributed averagly in space domain after the signal is implemented DFT. It enables the schemes further to embed the watermark with the magnitude of its coefficients.

Given a two-dimensional signal f(x, y), the DFT is defined

$$F(u,v) = \frac{1}{MN} \sum_{x=0}^{M-1} \sum_{y=0}^{N-1} f(x,y) e^{\left(-j2\pi(ux/M + vy/N)\right)} \qquad (7)$$

For u = 0, 1, 2..., M-1, v = 0, 1, 2 ,., N-1 and j=√-1

The inverse DFT (IDFT) is given by:

$$F(x,y) = \sum_{u=0}^{M-1} \sum_{v=0}^{N-1} F(u,v) e^{\left(j2\pi(ux/M + vy/N)\right)} \qquad (8)$$

where, (M, N) are the dimensions of the image.

The DFT is useful for watermarking purposes because it helps in selecting the adequate parts of the image for embedding, in order to achieve the highest invisibility and robustness.

1.3.2 The wavelets transform

Wavelet transform decomposes an image into a set of band limited components which can be reassembled to reconstruct the original image without error. The DWT (Discrete Wavelet Transform) divide the input image into four non-overlapping multi-resolution sub-bands

LL1, LH1, HL1 and HH1. The process can then be repeated to computes multiple "scale" wavelet decomposition, as in the 2 scale wavelet transform shown in Fig. 6.

One of the many advantages over the wavelet transform is that it is believed to more accurately model aspects of the HVS as compared to the FFT or DCT. This allows us to use higher energy watermarks in regions that the HVS is known to be less sensitive, such as the middle frequency bands (LH, HL) and high resolution band (HH). But watermark embedded in high resolution band can be easily be distorted by geometric transformation, compression and various signal processing operations.

Embedding watermarks in middle frequency regions allow us to increase the robustness of our watermark, at little to no additional impact on image quality [6].

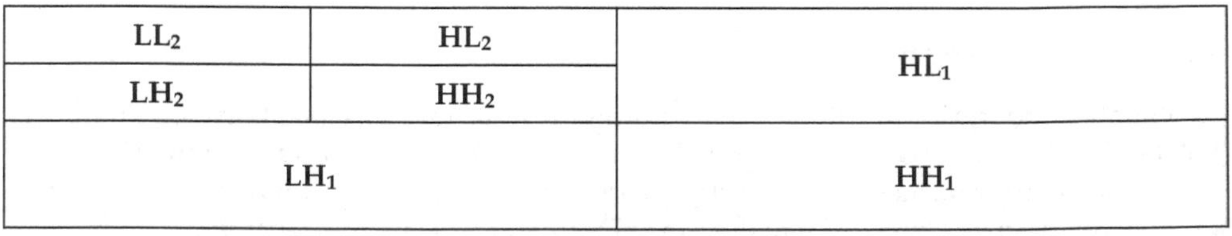

LL_2	HL_2	HL_1
LH_2	HH_2	
LH_1		HH_1

Fig. 6. Scale 2 Dimensional DWT

1.4 Watermarking applications

Although the main motivation behind the digital watermarking is the copyright protection, its applications are not that restricted. There is a wide application area of digital watermarking, including broadcast monitoring, fingerprinting, authentication and covet communication [7, 8, 9, 10].

By embedding watermarks into commercial advertisements, the advertisements can be monitored whether the advertisements are broadcasted at the correct instants by means of an automated system [7, 8]. The system receives the broadcast and searches these watermarks identifying where and when the advertisement is broadcasted. The same process can also be used for video and sound clips. Musicians and actors may request to ensure that they receive accurate royalties for broadcasts of their performances.

Fingerprinting is a novel approach to trace the source of illegal copies [7, 8]. The owner of the digital data may embed different watermarks in the copies of digital content customized for each recipient. In this manner, the owner can identify the customer by extracting the watermark in the case the data is supplied to third parties. The digital watermarking can also be used for authentication [7, 8]. The authentication is the detection of whether the content of the digital content has changed. As a solution, a fragile watermark embedded to the digital content indicates whether the data has been altered. If any tampering has occurred in the content, the same change will also occur on the watermark. It can also provide information about the part of the content that has been altered.

Covert communication is another possible application of digital watermarking [7,8]. The watermark, secret message, can be embedded imperceptibly to the digital image or video to communicate information from the sender to the intended receiver while maintaining low probability of intercept by other unintended receivers.

There are also non-secure applications of digital watermarking. It can be used for indexing of videos, movies and news items where markers and comments can be inserted by search engines [8]. Another non-secure application of watermarking is detection and concealment of image/video transmission errors [11]. For block based coded images, a summarizing data of every block is extracted and hidden to another block by data hiding. At the decoder side, this data is used to detect and conceal the block errors.

1.5 Watermarking requirements

The efficiency of a digital watermarking process is evaluated according to the properties of perceptual transparency, robustness, computational cost, bit rate of data embedding process, false positive rate, recovery of data with or without access to the original signal, the speed of embedding and retrieval process, the ability of the embedding and retrieval module to integrate into standard encoding and decoding process etc. 7, 8, 9, 12, 13].

Depending on the application, the properties, which are used mainly in the evaluation process, varies.

The main requirements for copyright protection are imperceptibility and robustness to intended or non-intended any signal operations and capacity.

The owner of the original data wants to prove his/her ownership in case the original data is copied, edited and used without permission of the owner. In the watermarking research world, this problem has been analyzed in a more detailed manner [13, 14, 15, 16, 17, 18].

The imperceptibility refers to the perceptual similarity between the original and watermarked data. The owner of the original data mostly does not tolerate any kind of degradations in his/her original data. Therefore, the original and watermarked data should be perceptually the same. Robustness to a signal processing operation refers to the ability to detect the watermark, after the watermarked data has passed through that signal processing operation.

The robustness of a watermarking scheme can vary from one operation to another. Although it is possible for a watermarking scheme to be robust to any signal compression operations, it may not be robust to geometric distortions such as cropping, rotation, translation etc. The signal processing operations, for which the watermarking scheme should be robust, changes from application to application as well. While, for the broadcast monitoring application, only the robustness to the transmission of the data in a channel is sufficient, this is not the case for copyright protection application of digital watermarking. For such a case, it is totally unknown through which signal processing operations the watermarked data will pass. Hence, the watermarking scheme should be robust to any possible signal processing operations, as long as the quality of the watermarked data preserved.

The capacity requirement of the watermarking scheme refers to be able to verify and distinguish between different watermarks with a low probability of error as the number of differently watermarked versions of an image increases [17]. While the robustness of the watermarking method increases, the capacity also increases where the imperceptibility decreases. There is a trade off between these requirements and this trade off should be taken into account while the watermarking method is being proposed.

2. Proposed techniques

We have proposed a blind invisible watermarking technique for copyright protection of the colored images.

In blind techniques, during the extraction process original image is not required. Watermarking systems which involve marking imperceptible alteration on the cover data to convey the hidden information, is called invisible watermarking. Here 512*512 grayscale image of 'peppers' is taken as host image and 32*32 binary image is taken as watermark image.

Then implement second level wavelet transform on host image using wavelet function 'haar' and extracted middle level components (HL_2, LH_2) for embedding watermark. Middle level components are selected for embedding watermark as much of the signal energy lies at low-frequencies sub-band which contains the most important visual parts of the image and high frequency components of the image are usually removed through compression and noise attacks.

We have divided the HL_2 and LH_2 bands in to 4x4 blocks and applied DFT in these blocks and used to two highly uncorrelated pseudo random sequences (treated as key: key1) to embed watermarking message according to template matrix. We reshaped watermark image of 32x32 into a row vector of size 1024, called watermark message.

A template matrix is such a matrix whose size is 4x4 and elements are 0 and 1. Watermarking message is embedded into image blocks only in the position where the template matrix's element is 1. Through amounts of experiments, we found when the template matrix is set to $[1,1,1,1;1,0,0,1;0,0,0,1;0,0,0,1]^T$,imperceptibility and robustness of the algorithm can get better balance. Here template matrix is used as a key2.

We have embedded watermark according to the (9) given below.

$$I_w(x, y) = I(x, y) + k * W(x, y) \qquad (9)$$

In (9), k denotes a gain factor, and I_W the resulting watermarked image, I the cover image and W the watermark to be embedded. Increasing k increases the robustness of the watermark at the expense of the quality of the watermarked image.

The algorithm for the proposed method is given below:

The watermark embedding steps are as follows:

- Implement second level wavelet transform on host Image H using wavelet function 'Haar' and Extract middle frequency components (LH_2, HL_2).
- Divide the HL_2, LH_2 components in several blocks of size 4x4 and DFT is applied to these blocks.
- Perform search to find highly uncorrelated pseudo random (PN) sequences (seq_zero and seq_one) and use these as a key1.
- Defines the template matrix of an 4x4.
- Set gain factor K and embed the watermark to the cover image under the following rule:
- If wa(i,j) ==0 then

```
        If template(m,n)==1 then
        I(m,n)=I(m,n) + K*seq_zero(m,n)
        End
        Else
        if template(m,n)==1 then
        I(m,n)=I(m,n)+K*seq_one(m,n)
        End
        End
```

Where $1 \le i \le M$, $1 \le j \le N$, and $1 \le m,n \le 4$

Here I denotes to 4x4 DFT blocks.

- Apply IFFT to each image block and use the result as the middle frequency component of DWT to recover the component which has been embedded watermarking messages.
- Replace the component of the host image by the watermarked component.
- Display watermarked image.

The watermark extraction steps of this technique are as follows:

- Implement Wavelet transform on Host image using wavelet function 'Haar' and Extract middle frequency components (LH_2, HL_2).
- Divide the HL_2, LH_2 components in several blocks of size 4x4 and DFT is applied to these blocks.
- Use same highly uncorrelated PN sequences (key1) and the template matrix of 4x4 (key2) to select elements that are embedded watermarking message to make up sequence.
- Calculate the correlation separately between sequence and seq_zero and between sequence and seq_one. The result is stored in corr_zero and corr_one respectively.
- Detect the watermark according the following rule:

```
        If corr_zero(i) > corr_one(i) then
        watermark_detected(i)=0;
        Else
        watermark_detected(i)=1
        End
```

- Reshape the recovered message.
- Display recovered message.
- Calculate the quality of recovered image by using PSNR function according to the (4).
- Calculate the Accuracy rate of recovered image by using AR function as per the (6).

3. Experimental results

In this section, we show some experimental results to demonstrate the effectiveness and success of our digital watermarking techniques. The standard 512 × 512 grayscale image "pepper" is used as host image, as shown in Fig. 7. The 32 × 32-pixels binary image is used as the watermark image, as shown in Fig. 8.

We applied the peak-signal to noise rate (PSNR) given in (4) to measure the image quality of an attacked image and accuracy rate AR given in (5) to evaluate the robustness of a copyright protection scheme for a specific attack.

Original Gray Image

Fig. 7. Original image of pepper

MCK

Fig. 8. Watermark image

3.1 Experimental result and analysis

The experimental results are represented in the following, respectively for watermarked image and extracted watermark image as shown in Fig. 9 (i), and Fig. 9 (ii), while taking the different values of gain factor K. And various observations for experiment are depicted in Table I.

K=10

K=20

K=30

K=50

Fig. 9. (i) Watermarked image, (ii) Extracted watermark image

Exhaustive testing against signal processing operation, Geometric distortion , collusion still has to be performed.

Gain Factor(K)	Execution Time	AR	PSNR
K=10	176.6796	89.4532	60.3456
K=20	175.6307	93.0640	55.2389
K=30	173.3867	91.3765	54.7829
K=40	175.3912	93.5621	54.6047

Table 1.

4. Conclusion

The need for digital watermarking on electronic distribution of copyright material is becoming more prevalent. In this paper an overview of the digital watermarking techniques are given and a blind invisible watermarking technique for grayscale images based on DWT and DFT is presented. The algorithm use 512*512 gray images as a host image and 32*32 binary image as watermarked image.

Firstly, two level wavelet decomposition is implemented on the host image. Then, the middle frequency components are extracted and divided in to several blocks of size 4*4 and DFT is implemented on them. Finally, two pseudo random sequences are created and embedded to blocks which have implemented DFT according to whether the corresponding position is 0 or 1 in the watermark matrix which has been implemented.

The original image is not required while extracting the watermark Instead, correlations among each block and two sequences are respectively calculated. Watermark is recovered on foundation of the relative magnitude of correlation between the corresponding block and one sequence or the other. The idea of applying two transform is based on the fact that combined transforms could compensate for the drawbacks of each other, resulting in effective watermarking.

5. References

[1] S.P. Mohanty, et al., "A Dual Watermarking Technique for Images", Proc.7th ACM International Multimedia Conference, ACM-MM'99, Part 2, pp 49-51, Orlando, USA, Oct. 1999

[2] Yusnita Yusof and Othman O. Khalifa , "Digital Watermarking For Digital Images Using Wavelet Transform", Proc 2007 IEEE conference, pp 665-669

[3] Potdar, V., S. Han and E. Chang, 2005. "A Survey of Digital Image Watermarking Techniques", in Proc. of the IEEE International Conference on Industrial Informatics, pp: 709-716, Perth, Australia.

[4] Chan, C. and L. Cheng, 2004. "Hiding Data in Images by Simple LSB Substitution", Pattern Recognition, 37(3):469-474.

[5] Wang, R., C. Lin and J. Lin, " Copyright protection of digital images by means of frequency domain watermarking," Proc. of the SPIE Conference On Mathematics of Data/Image Coding, Compression, and Encryption, USA.

[6] G. Langelaar, I. Setyawan, R.L. Lagendijk, "Watermarking Digital Image and Video Data", in IEEE Signal Processing Magazine, Vol 17, pp 20-43, September 2000.

[7] Ingemar J.Cox, Matt L. Miller and Jeffrey A. Bloom, "Watermarking Applications and their properties", Int. Conf. On Information Technology'2000, Las Vegas, 2000.

[8] Gerhard C. Langelaar, Iwan Setyawan, and Reginald L. Lagendijk, "Watermarking Digital Image and Video Data", IEEE Signal Processing Magazine, September 2000.

[9] Maurice Mass, Ton Kalker, Jean-Paul M.G Linnartz, Joop Talstra, Geert F. G. Depovere, and Jaap Haitsma, " Digital Watermarking for DVD Video Copy Protection", IEEE Signal Processing Magazine, September 2000.

[10] Fabien A.P. Petitcolas, " Watermarking Schemes Evaulation", IEEE Signal Processing Magazine, September 2000

[11] Technical Report, submitted to The Scientific and Technical Research Council of Turkey (Tübitak) under project EEEAG 101E007, April 2002

[12] Jean François Delaigle, " Protection of Intellectual Property of Images by perceptual Watermarking", Ph.D Thesis submitted for the degree of Doctor of Applied Sciences, Universite Catholique de Louvain, Belgique.

[13] Ingemar J. Cox, Joe Kilian, Tom Leighton, and Talal Shamoon, "Secure Spread Spectrum Watermarking for Multimedia", IEEE Trans. on Image Processing, 6, 12, 1673-1687, (1997).

[14] Mitchell D. Swanson, Mei Kobayashi, and Ahmed H. Tewfik, "Multimedia Data-Embedding and Watermarking Technologies", Proceedings of the IEEE., Vol. 86, No. 6, June 1998.

[15] Mitchell D. Swanson, Bin Zhu, and Ahmed H. Tewfik, "Transparent Robust Image Watermarking", 1996 SPIE Conf. on Visual Communications and Image Proc.

[16] Christine I. Podilchuk and Wenjun Zeng, " Image-Adaptive Watermarking Using Visual Models", IEEE Journal of Selected Areas in Communications, Vol.16, No.4, May 1998.

[17] Raymond B. Wolfgang, Christine I. Podilchuk and Edward J. Delp, "Perceptual Watermarks for Image and Video", Proceedings of the IEEE, Vol. 87, No. 7, July 1998.

[18] Sergio D. Servetto, Christine I. Podilchuk, Kannan Ramchandran, "Capacity Issues in Digital Image Watermarking", In the Proceedings of the IEEE International Conference on Image Processing (ICIP), Chicago, IL, October 1998.

7

The Wavelet Transform as a Classification Criterion Applied to Improve Compression of Hyperspectral Images

Daniel Acevedo and Ana Ruedin
Departamento de Computación
Facultad de Ciencias Exactas y Naturales
Universidad de Buenos Aires
Argentina

1. Introduction

Satellites continually feeding images to their base, pose a challenge as to the design of compression techniques to store these huge data volumes. We aim at lossless compression of hyperspectral images having around 200 bands, such as AVIRIS images. These images consist of several images (or bands) obtained by filtering radiation from the earth at different wavelengths. Compression is generally achieved through reduction of spatial as well as spectral correlations.

Most of the hyperspectral compressors are prediction–based. Since spectral correlation is usually high (much more higher than spatial correlation) pixels are predicted with other pixels in an adjacent band (rather than other pixels surrounding the one to be predicted). SLSQ (Rizzo et al., 2005), a low-complexity method designed for hyperspectral image compression, performs a simple prediction for each pixel, by taking a constant times the same pixel in the previous band. The constant is calculated by least squares over 4 previously encoded neighboring pixels. SLSQ–OPT version of SLSQ performs one band look–ahead to determine if the whole band is better compressed this way or with intraband prediction, while in the SLSQ–HEU version this decision is taken by an offline heuristic. CCAP (Wang et al., 2005) predicts a pixel with the conditional expected value of a pixel given the context. The expected value is calculated over coded pixels having matching (highly correlated) contexts. Slyz and Zhang (Slyz & Zhang, 2005) propose 2 compressors (BH and LM) for hyperspectral images. BH predicts a block as a scalar times the same block in the previous band. Coding contexts are defined by the quantized average error. LM predicts a pixel by choosing among different intraband predictions the one that works best for several pixels at the same position in previous bands.

Mielikainen and Toivanen proposed C-DPCM (Mielikainen & Toivanen, 2003), a method that classifies the pixels at the same location and through all the bands, with vector quantization. Interband prediction is performed using the pixels at the same position in 20 previous bands. Weights, calculated for each class/ band, are sent into the code, as well as the 2D template with the classes. Aiazzi et al. (Aiazzi et al., 1999) classify the prediction context of every pixel using fuzzy clustering, and then calculate the weights for each class. For compression of hyperspectral images (Aiazzi et al., 2007) they divide each band into small blocks (16×16), they calculate weights for interband prediction over each pixel in a block, and then make a fuzzy clustering of the weights obtained for all the blocks. For each pixel a membership degree is computed according to the efficiency of the weights (from different clusters) on a causal context of the pixel. The final prediction for a pixel is obtained by a combination of linear predictions involving weights from different clusters, pondered by the degrees of membership of the pixel to each cluster. It is worth mentioning that wavelet-based compressors such as JPEG2000 (Taubman & Marcellin, 2002) have been successfully used for lossy compression of multiband images, either hyperspectral (Blanes & Serra-Sagrista, 2010; Fowler & Rucker, 2007) or general earth data (Kulkarni et al., 2006).

In this work we will improve the well-known algorithm which was developed for hyperspectral images: LAIS-LUT. This algorithm makes predictions of each pixel using other pixels in the same band. Which pixel is used for prediction is determined by inspecting in the previously encoded band. This algorithm uses two possible candidates for prediction. We will introduce the wavelet transform as a tool for classification in order to make better decisions about which of these two possible candidates acts as a more appropriate prediction. First, LAIS-LUT will be introduced, as well as the LUT algorithm on which it is based. Then we show how the wavelet transform is used to improve it. Finally, we give some results and conclusions of our method, called Enhanced LAIS-LUT.

2. Look-up table based algorithms

2.1 LUT algorithm

The Look Up Table algorithm (Mielikainen, 2006) is a fast compression technique based on predicting a pixel with another pixel from the same band. The previous band is inspected in order to determine which pixel in the same band is used for prediction. As an example, suppose pixel $I_{x,y}^{(z)}$ in band z wants to be predicted. Then we seek on band $z-1$ the pixel with the same intensity as $I_{x,y}^{(z-1)}$ which is nearest to it in a causal neighborhood. Let $I_{x',y'}^{(z-1)}$ be that pixel. Then, the prediction for pixel $I_{x,y}^{(z)}$ will be $I_{x',y'}^{(z)}$. If no match is found, pixel $I_{x,y}^{(z-1)}$ is the one selected for prediction. In order to speed things up, a look up table data structure is used for searching the pixel on the previous band. With this data structure the algorithm is efficiently implemented as shown in Fig 1 for consecutive bands z and $z-1$. Then, the difference $I^{(z)} - P^{(z)}$ between the band and its prediction is entropy coded and this process is repeated for $z = 2, \ldots, 224$.

In Fig. 2 entropy values are ploted for each band of the Jasper Ridge image. In dashed line, entropies of the pixels of the image are ploted. 6 steps of the 2D S+P wavelet transform were computed and the entropy of the coefficients is plotted in dotted line for each band. Finally it is shown in gray line the entropy of the prediction differences for the LUT algorithm. It is remarkable how high is the compression achieved with this simple algorithm, which is only based on indexing and updating a table. It is entirely based on the premise of high correlation between bands and designed in order to take advantage of this fact.

Data: Bands $I^{(z)}$, $I^{(z-1)}$ and table initialized as LUTable$(i) = i$
Result: Prediction for band z: $P^{(z)}$
for *every pixel* $I_{x,y}^{(z)}$ **do**

$\quad\quad P_{x,y}^{(z)} \leftarrow$ LUTable$(I_{x,y}^{(z-1)})$;

$\quad\quad$ LUTable$(I_{x,y}^{(z-1)}) \leftarrow I_{x,y}^{(z)}$;

end

Fig. 1. Look-up Table algorithm.

Original	Wavelet S+P	LUT prediction
8.6656	6.6587	4.9504

Table 1. Average entropies for first scene of Jasper Ridge.

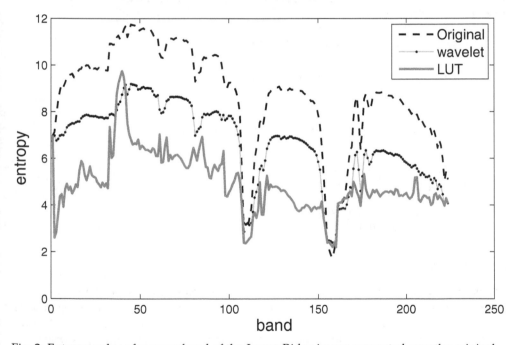

Fig. 2. Entropy values for every band of the Jasper Ridge image computed over the original image (dashed), over the prediction differences for the LUT method (gray), and over the 2D S+P wavelet transform (dotted). See averaged values in Table 1.

2.2 LAIS-LUT

An improvement over the LUT algorithm has been presented in (Huang & Sriraja, 2006). It was named LAIS-LUT after Locally Averaged Interband Scaling LUT and it behaves more accurately in presence of outliers. This modification adds an extra LUT table and the predictor is selected from one of the two LUTs. Using a scaling factor α which is precomputed on a causal neighbourhood, an estimate $\tilde{P}_{x,y}^{(z)}$ is calculated for a current pixel $I_{x,y}^{(z)}$ as $\tilde{P}_{x,y}^{(z)} = \alpha I_{x,y}^{(z-1)}$, where

$$\alpha = \frac{1}{3}\left(\frac{I_{i-1,j}^{(z)}}{I_{i-1,j}^{(z-1)}} + \frac{I_{i,j-1}^{(z)}}{I_{i,j-1}^{(z-1)}} + \frac{I_{i-1,j-1}^{(z)}}{I_{i-1,j-1}^{(z-1)}}\right) \tag{1}$$

Since two values are now possible candidates for prediction (one for each LUT), the one that is closer to $\tilde{P}_{x,y}^{(z)}$ is selected as the final prediction (Fig. 3 shows LAIS-LUT algorithm). When prediction $P^{(z)}$ for band z is estimated, the prediction error $I^{(z)} - P^{(z)}$ is entropy coded. This is repeated for $z = 2, \ldots, 224$. Notice that when the tables are not initialized, $P_{x,y}^{(z)}$ is selected for prediction, and, the value of the first LUT is used for prediction only when this table is initialized.

Data: Bands $I^{(z)}$, $I^{(z-1)}$, tables LUTable$_1$ and LUTable$_2$ are not initialized
Result: Prediction for band z: $P^{(z)}$
for *every pixel $I_{x,y}^{(z)}$* **do**
   if *both tables are not initialized on entry (x,y)* **then**
      $P_{x,y}^{(z)} \leftarrow \alpha I_{x,y}^{(z-1)}$;
   else if *only LUTable$_1$ is initialized in (x,y)* **then**
      $P_{x,y}^{(z)} \leftarrow \text{LUTable}_1(I_{x,y}^{(z-1)})$;
   else
      $P_{x,y}^{(z)} \leftarrow$ closer value of $\{\text{LUTable}_1(I_{x,y}^{(z-1)}), \text{LUTable}_2(I_{x,y}^{(z-1)})\}$ to $\alpha I_{x,y}^{(z-1)}$;
   end
   $\text{LUTable}_2(I_{x,y}^{(z-1)}) \leftarrow \text{LUTable}_1(I_{x,y}^{(z-1)})$;
   $\text{LUTable}_1(I_{x,y}^{(z-1)}) \leftarrow I_{x,y}^{(z)}$;
end

Fig. 3. LAIS-LUT algorithm.

2.3 LAIS-QLUT

Mielikainen and Toivanen proposed a modification to the LAIS-LUT method in order to shrink Lookup tables (Mielikainen & Toivanen, 2008). For that, the value used for indexing into the LUTs is quantized and smaller LUTs can be used (i.e., the value x used as an index in the LUT is replaced by $\lfloor x/q \rfloor$ for some quantization step q, being $\lfloor \cdot \rfloor : \mathbb{R} \rightarrow \mathbb{Z}$ a function that maps a real number to a close integer). In the previous section, LAIS-LUT found exact matches in the previous band in order to obtain the predictor in the current band. For LAIS-QLUT, this search is no more exact and the predictor's selection in the current band is based on 'similarities' in the previous band. The best quantization step is obtained by an exhaustive search on each band (LAIS-QLUT-OPT version) or determined offline for each band by training on a set of images (LAIS-QLUT-HEU version). We decided not to add classes to the context over LAIS-QLUT, since a combination of the two would result in a prohibitive increase in complexity.

3. The wavelet transform as a tool for classification

In order to improve compression bitrates, the scaling factor in LAIS-LUT will be estimated more accurately. For that, a classification stage will be added to the algorithm. The scaling factor is computed as an average of quotients of collocated pixels in consecutive bands according to Equation 1. These pixels belong to a close neighbor of the pixel to be predicted (either on the current band, or on the previous band). Once classes are established for each pixel, not all the pixels in the close causal neighbor will be used for estimating the scaling factor α. Instead, we may use only those pixels that belong to the same class of the pixel to be predicted so as to obtain a more accurate estimation. This will enhance the LAIS-LUT algorithm, by allowing a better decision on which of the two look up tables yields a better prediction.

In order to establish classes we will make use of the wavelet transform. Since hyperspectral images have a considerable number of bands, the wavelet transform can be applied in the spectral direction. AVIRIS images have 224 bands. Considering each pixel as a vector in $\mathbb{Z}^{224}$, each of them may be 1D-wavelet transformed. And with the information of the wavelet transform of each 'pixel' classes can be determined. First, the wavelet transforms used in this work are introduced.

3.1 The wavelet transform

The wavelet transform allows the representation of a signal in multiresolution spaces. In the wavelet representation, the transformed signal can be viewed as an approximation of the original signal plus a sum of details at different levels of resolution. Each of these details and approximations are associated to function basis which have good time-frequency localization (Mallat, 1999). In images –a simple extension of the 1-dimensional case–, decorrelation is achieved obtaining a sparse representation. Two different types have been considered in this work: orthogonal wavelets and lifting-based wavelets.

For the classical orthogonal wavelet, consider a 1D signal $x = [x_n]_{n=0,...,N-1}$ of length N (even). For a step of the wavelet, x is transformed into approximation coefficients $s = [s_n]_{n=0,...,\frac{N}{2}-1}$ and detail coefficients $d = [s_n]_{n=0,...,\frac{N}{2}-1}$, where s is the result of convolving x with a lowpass filter followed by a decimation. The same process happens to d, but a high pass filter is used instead. For more steps, the wavelet transform is applied recursively over the approximation coefficients s. The original signal can be recovered if the inverse process is carried out in order (upsampling followed by the convolution with the corresponding reversed filter –see Fig. 4). Depending on the filter, wavelets with different properties can be obtained. In this work we use the Daubechies 4 wavelet (Daubechies, 1992) whose lowpass filter is $[h_3, h_2, h_1, h_0] = \frac{1}{4\sqrt{2}}[3+e, 1-e, 3-e, 1+e]$ with $e = \sqrt{3}$, and high pass filter $g' = [-h_0, h_1, -h_2, h_3]$. The Symmlet wavelet, with 8 vanishing moments (and filter of length 16), was also used in this work.

Another way of constructing different wavelets is by the lifting scheme (Daubechies & Sweldens, 1998). They are built in the spatial domain. The basic idea is to split the signal into two components, for instance, odd samples ($x_{odd} = x_{2k+1}$) and even samples ($x_{even} = x_{2k}$). Then, detail coefficients are obtained by using one component to predict the other: $d = x_{odd} - P(x_{even})$. Better predictions will yield more zeros, and therefore more decorrelation is

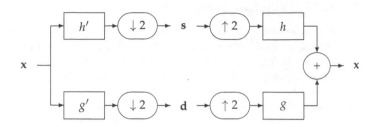

Fig. 4. Orthogonal wavelet analysis via convolutions and decimations followed by the synthesis via upsampling and convolution. Highpass filter g is obtained from lowpass filter h as $g_k = (-1)^k h_{1-k}$, and g' indicates the g filter reversed.

achieved. In order to obtain the approximation part of the transformed signal, the unpredicted component is softened with an 'update' of the detail previously obtained: $\mathbf{s} = \mathbf{x}_{\text{even}} + U(\mathbf{d})$. See Fig. 5 for an illustration of this scheme. With the lifting scheme it is possible to construct wavelets that map integers into integers (Calderbank et al., 1998) by the use of a rounding operator at the prediction or update stage. For this work the $(2,2)$ and S+P wavelets have been considered (see Table 2).

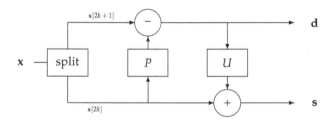

Fig. 5. Lifting scheme analysis for a 1-D signal $\mathbf{x}$.

Wavelet	Formula
$(2,2)$	$d_n = x_{2n+1} - \left\lfloor \frac{1}{2}(x_{2n} + x_{2n+2}) + \frac{1}{2} \right\rfloor$
	$s_n = x_{2n} + \left\lfloor \frac{1}{4}(d_{n-1} + d_n) + \frac{1}{2} \right\rfloor$
S+P	$d_n^{(1)} = x_{2n+1} - x_{2n}$
	$s_n = x_{2n} + \left\lfloor \frac{d_n^{(1)}}{2} \right\rfloor$
	$d_n = d_n^{(1)} + \left\lfloor \frac{2}{8}(s_{n-1} - s_n) + \frac{3}{8}(s_n - s_{n+1}) + \frac{2}{8}d_{n+1}^{(1)} + \frac{1}{2} \right\rfloor$

Table 2. Wavelet transforms that maps integers into integers.

4. Classification of hyperspectral images

Using images captured by the scanner system called AVIRIS (Airborne Visible/Infrared Imaging Spectrometer) developed by JPL (Jet Propulsion Laboratory; see http://aviris.jpl.nasa.gov), we aim at determining classes for pixels. These images have 224 bands, each corresponding to a response in a certain range of the electromagnetic spectrum. Each pixel represents 20 meters and is allocated in a 2 byte signed integer. AVIRIS images usually have 614 columns and every 512 rows, the image is partitioned into 'scenes', and each scene is stored in a different file.

Considering each pixel $I_{x,y}$ as a vector in $\mathbb{Z}^{224}$ where each component belongs to a different band, different behaviours depending on what type of soil is being considered can be observed. Figure 6 shows the first scene of the Jasper Ridge image with four pixels belonging to different classes, marked with symbols '$\oplus$', '$\circ$', '$\heartsuit$' and '$\diamondsuit$'. The spectral vectors associated to the same 4 positions are plotted in Figure 7; where the characteristic signal -called the spectral signature- of the type of soil can be observed.

Fig. 6. Band 50 of Jasper Rigde image.

In order to classify the image, for every pixel, a 1-D wavelet transform is applied along the spectral direction. The entropy of each transformed spectral vector is computed, giving an image C where $C_{x,y} := H(W(I_{x,y}))$, being $W(\cdot)$ a 1-D wavelet transform function and $H(\cdot)$ the entropy function weighted by wavelet subband. Since AVIRIS images have 224 bands, 5 steps of the wavelet transform are applied to each pixel. So we have 5 detail subbands and 1 approximation subband. Since decimation is performed after each step, these subbands are different in size. Therefore, the total entropy of the 1-D wavelet transform is computed as the

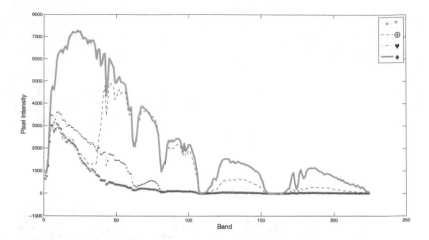

Fig. 7. Spectral signatures of different classes: values of selected pixels in Fig. 6 throughout the bands.

Fig. 8. Entropy of the wavelet transform of every pixel. Darker pixels indicate lower entropy values.

Image	Enhanced LAIS-LUT				LAIS-LUT	LUT
	Daubechies 4	Symmlet 8	S+P	(2,2)		
Low Altitude	3.17875	3.17546	3.17793	3.17797		
Jasper Ridge	3.43895	3.43855	3.43914	3.43833	3.42	3.23
Cuprite	3.71393	3.71394	3.71390	3.71390	3.58	3.44
Lunar Lake	3.58201	3.58198	3.58198	3.58201	3.53	3.4

Table 3. Compression ratios for Enhanced LAIS-LUT (with classes for estimating the scaling factor) and classical LAIS-LUT and LUT algorithms.

sum of the entropies of each subband, weighted by the size of the subband relative to the size of the whole transform.

In Figure 8 a grayscale image of the entropies is displayed. This image may be further split into classes with an unsupervised classifier. In this work, mean-shift (Comaniciu & Meer, 2002) was used.

5. Results

Results of compressing AVIRIS images with LAIS-LUT and the enhanced version (in which the scaling factor is calculated with pixels belonging to the same class, considering a causal neighbor of 4×4 pixels around the one to be predicted) is shown in Table 3.

The name of the wavelet in the table indicates the wavelet used to transform each pixel (and all the bands) followed by the entropy estimation and mean shift classification. Compression ratio results for LAIS-LUT and LUT algorithms were obtained from (Mielikainen & Toivanen, 2008). It can be observed that the enhanced version of LAIS-LUT using Daubechies 4 for classification outperforms the other methods.

We may conclude that the scaling factor α plays an important role in the compression performance of LAIS-LUT algorithm. When introduced, it was intended to decrease the deterioration produced by outliers in the original LUT algorithm. We have also been able to make use of the information in the wavelet domain and apply it to develop an efficient classifier. Since hyperspectral images have many bands because of their high spectral resolution, the information of the signal that each pixel represents (in all bands) was well captured by the wavelet transform and was fed into a powerful classifier such as mean-shift, giving good compression results.

6. References

Aiazzi, B., Alba, P., Alparone, L. & Baronti, S. (1999). Lossless compression of multi/hyperspectral imagery based on a 3-D fuzzy prediction, *IEEE Trans. Geos. Remote Sensing* 37(5): 2287–2294.

Aiazzi, B., Alparone, L., Baronti, S. & Lastri, C. (2007). Crisp and fuzzy adaptive spectral predictions for lossless and near-lossless compression of hyperspectral imagery, *IEEE Geos. Remote Sensing Letters* 4(4): 532–536.

Blanes, I. & Serra-Sagrista, J. (2010). Cost and scalability improvements to the karhunen-loêve transform for remote-sensing image coding, *Geoscience and Remote Sensing, IEEE Transactions on* 48(7): 2854 –2863.

Calderbank, A. R., Daubechies, I., Sweldens, W. & Yeo, B. (1998). Wavelet transforms that map integer to integers, *Applied and Computational Harmonics Analysis* 5(3): 332–369.

Comaniciu, D. & Meer, P. (2002). Mean shift: a robust approach toward feature space analysis, *IEEE Transactions on Pattern Analysis and Machine Intelligence* 24(5): 603–619.

Daubechies, I. (1992). *Ten lectures on wavelets*, Soc. Indus. Appl. Math.

Daubechies, I. & Sweldens, W. (1998). Factoring wavelet and subband transforms into lifting steps, *The Journal of Fourier Analysis and Applications* 4: 247–269.

Fowler, J. E. & Rucker, J. T. (2007). 3d wavelet-based compression of hyperspectral imagery, *in* C.-I. Chang (ed.), *Hyperspectral Data Exploitation: Theory and Applications*, John Wiley & Sons, Inc., pp. 379–407.

Huang, B. & Sriraja, Y. (2006). Lossless compression of hyperspectral imagery via lookup tables with predictor selection, *SPIE* 6365.

Kulkarni, P., Bilgin, A., Marcellin, M., Dagher, J., Kasner, J., Flohr, T. & Rountree, J. (2006). Compression of earth science data with jpeg2000, *in* G. Motta, F. Rizzo & J. A. Storer (eds), *Hyperspectral Data Compression*, Springer US, pp. 347–378.

Mallat, S. (1999). *A Wavelet Tour of Signal Processing*, Academic Press.

Mielikainen, J. (2006). Lossless compression of hyperspectral images using lookup tables, *IEEE SPL* 13(3): 157–160.

Mielikainen, J. & Toivanen, P. (2003). Clustered DPCM for the lossless compression of hyperspectral images, *IEEE Trans. Geos. Remote Sensing* 41(12): 2943–2946.

Mielikainen, J. & Toivanen, P. (2008). Lossless compression of hyperspectral images using a quantized index to look-up tables, *IEEE Geos. Remote Sensing Letters* 5(3).

Rizzo, F., Carpentieri, B., Motta, G. & Storer, J. A. (2005). Low complexity lossless compression of hyperspectral imagery via linear prediction, *IEEE Signal Processing Letters* 12(2): 138–141.

Slyz, M. & Zhang, L. (2005). A block-based inter-band lossless hyperspectral image compressor, *IEEE Proc. Data Compression Conf.* .

Taubman, D. & Marcellin, M. (2002). *JPEG2000: Image compression fundamentals, standards and practice*, Kluwer Academic Publishers, Boston.

Wang, H., Babacan, D. & Sayood, K. (2005). Lossless hyperspectral image compression using context-based conditional averages, *IEEE Proc. Data Compression Conference* .

Part 2

Applications in Engineering

8

Optimized Scalable Wavelet-Based Codec Designs for Semi-Regular 3D Meshes

Shahid M. Satti, Leon Denis, Ruxandra Florea,
Jan Cornelis, Peter Schelkens and Adrian Munteanu
Department of Electronics and Informatics (ETRO)
Vrije Universiteit Brussel-IBBT, Brussels,
Belgium

1. Introduction

3D graphics applications make use of polygonal 3D meshes for object's shape representation. The recent introduction of high-performance laser scanners and fast microcomputer systems gave rise to high-definition graphics applications. In such applications, objects with complex textures are represented using dense 3D meshes which consist of hundreds of thousands of vertices. Due to their enormous data size, such highly-detailed 3D meshes are rather intricate to store, costly to transmit via bandwidth-limited transmission media, and hard to display on end-user terminals with diverse display capabilities. Scalable compression, wherein the source representation can be adapted to the users' requests, available bandwidth and computational capabilities, is thus of paramount importance in order to make efficient use of the available resources to process, store and transmit high-resolution meshes.

State-of-the-art scalable mesh compression systems can be divided into two main categories. A first category includes codecs that directly compress the irregular topology meshes in the spatial domain. In such codecs, the connectivity information is encoded losslessly while mesh simplification methods such as vertex coalescing (Rossignac & Borrel, 1993), edge decimation (Soucy & Laurendeau, 1996) and edge collapsing (Ronfard & Rossignac, 1996) are employed to encode geometry. These mesh simplification methods progressively remove those mesh vertices which yield the smallest distortion. In order to enable the reconstruction of the original mesh at various levels of detail (LODs), the discarded vertices are encoded in the compressed bit-stream. Mesh compression systems belonging to this category include Progressive Meshes (Li & Kuo, 1998), (Pajarola & Rossignac, 2000) and Topological Surgery (Taubin et al., 1998). These techniques generally exhibit two major drawbacks: first, due to the highly irregular topology of the input mesh, a large source rate is needed for lossless encoding of connectivity. Secondly, encoding the removed vertices in the compressed bit-stream is quite costly for high-resolution meshes. Therefore, such schemes are not useful for complex meshes containing a large number of vertices. An alternative that solves the problem of the large source rates needed to encode the connectivity information, described above, is remeshing, which can be used to convert the original irregular mesh into a mesh consisting of regular elements, such as B-spline (Eck & Hoppe, 1996) or subdivision connectivity patches (Eck et al., 1995). The regular mesh lends itself better to compression, and hence compared to the irregular mesh a much lower rate is needed to losslessly encode its connectivity information. Furthermore, multiresolution techniques alleviate the second problem of having to encode all the original vertices, because only detail information has to be encoded in order to create multiple LODs (or multiple resolution levels). Remeshing together with subdivision-based multiresolution (Lounsbery et al., 1997) are the two major components of the second category of codecs which use space-frequency dilation methods such as wavelet transforms to decorrelate the input mesh data (Khodakovsky et al., 2000), (Denis et al., 2010b). The generated wavelet coefficients are compressed using tree-based bit-plane coding methods (Shapiro, 1993), (Munteanu et al., 1999b) to achieve high compression efficiency. Multiresolution mesh compression techniques provide substantial compression gains compared to their competing schemes, and in this chapter we will confine our discussion to these techniques only.

In the recent past, several multiresolution scalable mesh compression schemes have been proposed. The majority of these schemes use coding techniques which were specifically developed for image compression. However, in general, image and mesh data exhibit different statistical characteristics as the images are consisting of pixels (with intensities) while mesh data involve geometry, i.e., the positions of vertices in a 3D space. Thus, one must be cautious when extrapolating image compression techniques towards mesh geometry encoding.

In this book chapter, we propose a constructive design methodology for multiresolution-scalable mesh compression systems. The input mesh is assumed to possess subdivision connectivity (Lounsbery et al., 1997), i.e., the connectivity in the mesh is built through subdivision[1]. A 3D mesh with subdivision connectivity is also referred to as a semi-regular mesh. With respect to *design*, we address two major aspects of scalable wavelet-based mesh compression systems, namely, the optimality of embedded quantization in scalable mesh coding and the type of coefficient dependencies that can assure the best compression performance. In this context, thorough analyses investigating the aforementioned aspects are carried out to establish the most appropriate design choices. Later on, the derived design choices are integrated as components of the scalable mesh coding system to achieve state-of-the-art compression performance.

The remainder of the book chapter is organized as follows: in Section 2, a brief overview of multiresolution analysis of the mesh geometry is given. Section 3 presents a model-based theoretical investigation of optimal embedded quantization in wavelet-based mesh coding. An information theoretic analysis of the statistical dependencies among wavelet coefficients and the conclusions regarding the best exploitable statistical dependency are detailed in Section 4. Section 5 gives an overview of the state-of-the-art mesh compression systems.

2. Multiresolution analysis of semi-regular meshes

A 3D mesh $M = \{ \mathbf{c}, \mathbf{p} \}$ is generally represented as a set of two components, a vertex list $\mathbf{c}$ and a polygon list $\mathbf{p}$. $\mathbf{c}$ is a matrix whose ith row c_i contains the x, y and z position of the ith vertex, i.e., $c_i = \left[c_{i,x}, c_{i,y}, c_{i,z} \right]$. $\mathbf{p}$ is a list of polygons made up of edges where each edge is a line connecting two vertices. In computer graphics, 3D meshes are constructed

---

[1] In general, an initial remeshing step (Eck et al., 1995) is required to convert the original irregular mesh into a mesh with the required connectivity.

using different polygonal shapes, e.g., triangles, rectangles etc. However, in this chapter, we will confine our discussion to the triangular meshes only.

In the following, a brief theoretical overview of semi-regular multiresolution analysis is presented. Later on, two practical transforms, namely the lifting-based wavelet transform and the spatially adapted wavelet transform are detailed.

2.1 Theory

2.1.1 Subdivision surfaces

Subdivision is a process of iteratively refining a control polyhedron M^0 into fine geometry polyhedra such that the refined polyhedra $M^1, M^2, M^3 \ldots$ converge to a limit surface M^∞. In general, subdivision schemes consist of *splitting* and *averaging* steps. In the *splitting* step, each triangular face is split into four sub-triangles by adding new vertices. This way, an intermediate polyhedron $\bar{M}^j$ is created for any level j. The *averaging* step is used to determine the position of each vertex in M^j from its local neighborhood of vertices in $\bar{M}^j$, $j = 1, 2, \ldots, J$.

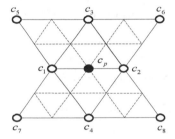

Fig. 1. Butterfly subdivision stencil.

$\mathbf{P}^j \in \mathbb{R}^{N_{j+1} \times N_j}$ and $\mathbf{Q}^j \in \mathbb{R}^{N_{j+1} \times N_{j+1}}$ (where N_j denotes the number of vertices of M^j) are the splitting and averaging matrix at level j. The subdivision process, expressed in matrix form, can be written as:

$$c^{j+1} = Q^j \cdot P^j \cdot c^j, \quad j = 0, 1, 2, \ldots, J-1.$$

A commonly used subdivision is Butterfly subdivision (Dyn et al., 1990). The subdivision stencil for Butterfly is shown in Fig. 1, where the position of a newly introduced vertex p is computed as, $c_p = \sum_{i=1}^8 a_i c_i$ whereby a_i's denote the Butterfly weights (Dyn et al., 1990). Loop (Loop et al., 2009) and Catmull-Clark (Catmull & Clark, 1978) are among the other commonly used subdivision schemes for 3D meshes.

2.1.2 Multiresolution analysis

Lounsbery (Lounsbery et al., 1997) first invented the multiresolution analysis for arbitrary topology semi-regular surfaces using subdivision. He proved that refinable bases exist when a coarse mesh M^0 is refined through subdivision, i.e.,

$$\phi^j(\mathbf{x}) = \phi^{j+1}(\mathbf{x}) \cdot \mathbf{P}^j \text{, for } \mathbf{x} \in M^0 \text{ and } 0 \le j < J. \tag{1}$$

$\phi^j(\mathbf{x})$ in the above equation denotes the row vector of scaling functions ϕ_i^j. Given these refinable scaling functions, scalar-valued function spaces associated with the coarsest geometry M^0 are defined as (Lounsbery et al., 1997):

$$V^j(M^0) := \mathrm{Span}(\phi^j(\mathbf{x})) \text{, for } 0 \le j < J. \tag{2}$$

Eq (1) implies that these spaces are indeed nested, i.e.,

$$V^0(M^0) \subset V^1(M^0) \subset V^2(M^0) \subset \dots, \tag{3}$$

The wavelet space $W^j(M^0)$ is defined as a space which is the orthogonal complement of $V^j(M^0)$ in $V^{j+1}(M^0)$. Hence, $W^j(M^0)$ and $V^j(M^0)$ together can represent any scalar-valued piecewise function in the space $V^{j+1}(M^0)$. If $\psi^j(\mathbf{x})$ is a row vector containing refinable bases functions of $W^j(M^0)$, the following stands (Lounsbery et al., 1997):

$$\psi^j(\mathbf{x}) = \phi^{j+1}(\mathbf{x}) \cdot \mathbf{Q}^j \text{, for } \mathbf{x} \in M^0 \text{ and } 0 \le j < J. \tag{4}$$

Combining (1) with (4) yields

$$\left(\phi^j(\mathbf{x}), \psi^j(\mathbf{x})\right) = \phi^{j+1}(\mathbf{x}) \cdot \left(\mathbf{P}^j, \mathbf{Q}^j\right) \text{, or } \left(\phi^j(\mathbf{x}), \psi^j(\mathbf{x})\right) \cdot \left(\mathbf{P}^j, \mathbf{Q}^j\right)^{-1} = \phi^{j+1}(\mathbf{x}). \tag{5}$$

A set of scaling functions $\phi^{j+1}(\mathbf{x})$ can then be used to decompose a surface S^{j+1} in $V^{j+1}(M^0)$, i.e.,

$$S^{j+1} = \sum_i c_i^{j+1} \phi_i^{j+1} = \phi^{j+1}(\mathbf{x}) \cdot \mathbf{c}^{j+1}, \tag{6}$$

where c_i^{j+1} is the ith vertex in M^{j+1}. Since the analysis filters are uniquely determined by the relationship

$$\left(\mathbf{P}^j \big| \mathbf{Q}^j\right)^{-1} = \left(\frac{\mathbf{A}^j}{\mathbf{B}^j}\right), \tag{7}$$

by combining Eq (5) and Eq (6) and making the above substitution for $\left(\mathbf{P}^j \big| \mathbf{Q}^j\right)^{-1}$, we obtain:

$$S^{j+1} = \phi^j(\mathbf{x}) \cdot \mathbf{A}^j \cdot \mathbf{c}^{j+1} + \psi^j(\mathbf{x}) \cdot \mathbf{B}^j \cdot \mathbf{c}^{j+1}. \tag{8}$$

From Eq (8), Eq (9) one derives the forward wavelet transform, given by:

$$\mathbf{c}^j = \mathbf{A}^j \cdot \mathbf{c}^{j+1}, \quad \mathbf{d}^j = \mathbf{B}^j \cdot \mathbf{c}^{j+1}, \quad \forall j: 0 \le j < J, \tag{9}$$

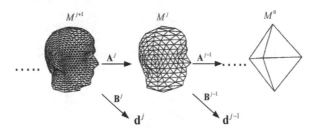

Fig. 2. Pictorial representation of the forward wavelet decomposition, (Lounsbery et al., 1997).

where $\mathbf{d}^j$ is a matrix containing the wavelet coefficients for the jth level of the transform. In general, after the transform, a fair amount of correlation still exist between x, y and z wavelet component. Local frame representation (Khodakovsky et al., 2000) of wavelet coefficients is often used to make wavelet components much more independent. After the local frame transformation, each wavelet coefficient consists of a *normal* and two *tangential* components. $\mathbf{A}^j$s and $\mathbf{B}^j$s are matrices representing the low and the high-pass filters, respectively, also referred to as analysis filter pairs.

A similar reasoning as for Eq (9) can be used to formulate the inverse wavelet transform, expressed by:

$$c^{j+1} = \mathbf{P}^j \cdot \mathbf{c}^j + \mathbf{Q}^j \cdot \mathbf{d}^j \text{, for } \forall j: \ 0 \le j < J. \tag{10}$$

Hence, $\mathbf{P}^j$s and $\mathbf{Q}^j$s jointly form the synthesis part of the decomposition for the lossless reconstruction of the input semi-regular mesh M^J. Note that the computation of the $\mathbf{A}^j$s and $\mathbf{B}^j$s involves the inversion of a large matrix, which makes the forward transform more complex than the inverse transform.

2.2 Lifting-based wavelet transform

As explained earlier, the filter bank implementation of multiresolution analysis is quite complex in the sense that the computation of analysis filters involve the computationally intensive inversion of large subdivision matrices. In this context, the lifting-based wavelet implementation (Schröder & Sweldens, 1995) provides a low complexity construction of multiresolution methods. In lifting-based multiresolution analysis, each scaling function ϕ_i^j of the jth level exists so that $\{\phi_i^j \mid i \in M^j\}$ is a Riesz basis of $V^j\left(M^0\right)$ (Schröder & Sweldens, 1995). The refinement relation for the scaling functions is then:

$$\phi_i^j = \sum_l p_{i,l}^j \cdot \phi_i^{j+1}, \tag{11}$$

where l is the set which defines all linear combination of scaling functions and $p_{i,l}^j$ forms the entries of a matrix similar to $\mathbf{P}^j$. A similar refinement relation as Eq (11) is also defined for wavelet functions, i.e., each wavelet function ψ_k^j exists so that $\{\psi_k^j \mid k \in K^j\}$ is a Riesz basis of $W^j\left(M^0\right)$:

$$\psi_k{}^j = \sum_l q_{k,l}^j \cdot \phi_k{}^{j+1}. \tag{12}$$

K^j and M^j are disjoint sets and they jointly form the scaling function index set of the next higher level, i.e., $M^{j+1} = M^j \oplus K^j$. The lifting-based forward decomposition is expressed by the following relations (Schröder & Sweldens, 1995):

$$\forall i \in M^j : \quad c_i^j = c_i^{j+1} \Rightarrow \text{subsample}$$

$$\forall k \in K^j : \quad d_k^j = c_k^{j+1} - \sum_{i \in M^j} a_i . c_i^j \Rightarrow \text{prediction} \tag{13}$$

$$\forall k \in K^j : \begin{cases} c_1^j = c_1^j + \tilde{a}_1 \cdot d_k^j \\ c_2^j = c_2^j + \tilde{a}_2 \cdot d_k^j \end{cases} \Rightarrow \text{update}$$

In the forward transform, the first step is to produce a lower-resolution mesh M^j starting from a higher-resolution version M^{j+1}. The wavelet coefficient d_k^j is the *prediction error* when a high-resolution vertex c_k^{j+1} is predicted based on its low-resolution neighborhood in M^j. After the prediction, an *update* step is used to modify the low resolution mesh M^j. The *update* step is carried out on a pair $\{c_1, c_2\}$ of low-resolution vertices joined by a parent edge (Schröder & Sweldens, 1995) using the update weights $\{\tilde{a}_1, \tilde{a}_2\}$. In general, the prediction and update weights only depend on the connectivity with respect to the vertex to be predicted. However, specific multiresolution analyses for which the weights depend on the specific resolution level and the underlying geometry can be also constructed (more details are given in Section 2.3).

The inverse transform can be formulated by following the forward-transform steps in the reverse order, i.e.:

$$\forall k \in K^j : \begin{cases} c_1^j = c_1^j - \tilde{a}_1 \cdot d_k^j \\ c_2^j = c_2^j - \tilde{a}_2 \cdot d_k^j \end{cases} \Rightarrow \text{inverse update}$$

$$\forall k \in K^j : \quad c_k^{j+1} = d_k^j + \sum_{i \in M^j} a_i \cdot c_i^j \Rightarrow \text{inverse predict} \tag{14}$$

$$\forall i \in M^j : \quad c_i^{j+1} = c_i^j \Rightarrow \text{inverse subsample}$$

2.3 Spatially Adaptive Wavelet Transform (SAWT)

As mentioned earlier, lifting-based transforms generally employ fixed prediction weights, independent of the spatial position and geometry around the vertex to be predicted. A simple observation reveals that a better prediction can result from adapting the prediction to the underlying geometry of the mesh. This argument is explained with a simple example: Fig. 3, referring to the position variable of the vertices, shows a scenario where the vertex to be predicted c_p lies on the straight line joining the vertex pair $\{c_1, c_2\}$, while the remaining six coarser vertices $\{c_i\}_{i=3}^8$ lie on two different planes. In this situation, a prediction function for c_p involving all eight coarser vertices will not be optimal and a better prediction could result by using c_1 and c_2 only. This is logical since c_p lies on the edge formed by the vertex pair $\{c_1, c_2\}$ and is geometrically more correlated to vertices $\{c_1, c_2\}$. Thus, an efficient prediction can be achieved if the prediction process is adapted to the local mesh geometry. Efficient prediction results in smaller energy of wavelet coefficients and hence an improved compression performance of the mesh coding system. To reverse the prediction operation, the decoder needs to know the weights used by the encoder for the prediction of each vertex c_p. Since additional rate (compared to classical Butterfly) needs to be spent for coding the prediction weights, the total compression efficiency in the geometry adaptive case is a compromise between the bitrate saved due to the efficient prediction and the extra bitrate needed for signaling the prediction weights.

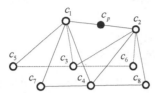

Fig. 3. Butterfly footprint on an edge.

In the following, a finite set of prediction filters is proposed in the context of spatially-adaptive wavelet transforms (SAWT) (Denis et al., 2010a). The idea is to use one filter out of this set which best suits the geometry around the vertex to be predicted and which results in the smallest prediction error. A careful application of such an adaptive approach will provide an average rate gain if the reduction in the bitrate due to better prediction dominates the extra bitrate needed to signal the filter type to the decoder.

In a first step, the input semi-regular mesh is segmented into regions as follows. Let $B(r,s)$ denote the bounding box of the input semi-regular mesh, where $r=(x_B,y_B,z_B)$ and $s=(s_x,s_y,s_z)$ represent the coordinates of the top-left corner and the size vector, respectively. Considering the bounding box as the root cell, each cell on a certain tree level is recursively split into eight equally sized sub-cells to create the next level of the octree. This recursive splitting continues until the number of vertices in the highest-level cells are smaller than a user-defined threshold α. This way, the semi-regular mesh is divided into regions of approximately the same size – see Fig. 4.

For each region k, the wavelet analysis is performed by selecting one of the six candidates filters given below:

$$f_1 = \tfrac{1}{2}(c_1+c_2)+\tfrac{1}{8}(c_3+c_4)-\tfrac{1}{16}(c_5+c_6+c_7+c_8) \Rightarrow (\text{Butterfly})$$
$$f_2 = \tfrac{1}{2}(c_1+c_2)+\tfrac{1}{4}(c_3+c_4)-\tfrac{1}{8}(c_5+c_6+c_7+c_8) \Rightarrow (\text{Modified Butterfly})$$
$$f_3 = \tfrac{3}{8}(c_1+c_2)+\tfrac{1}{8}(c_3+c_4) \Rightarrow (\text{Loop})$$
$$f_4 = \tfrac{1}{2}(c_1+c_2) \Rightarrow (\text{edge})$$
$$f_5 = \tfrac{1}{2}(c_3+c_4) \Rightarrow (\text{anti edge})$$
$$f_6 = \tfrac{1}{4}(c_1+c_2+c_3+c_4) \Rightarrow (\text{Hybrid of } f_4 \text{ and } f_5)$$

Note that the above set of filters is defined using a mixture of Butterfly, Loop and midpoint subdivision schemes.

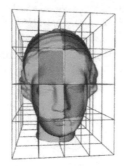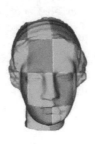

Fig. 4. Mesh partitioning for $\alpha = 400$. The red and green patches indicate different regions for which different prediction filters will be selected.

Similar to (Chang & Girod, 2006), a filter candidate for a particular region k in M^j, is chosen in an optimal distortion-rate (D-R) manner. More specifically, a predictor for each region k is selected such that the following Lagrangian cost function is minimized:

$$\Lambda_{M^j,k} = \underset{l \in \{1,2,..,6\}}{\arg\min} \left\{ E_k \left[\left(c_p - \tilde{c}_{p,f_l} \right)^2 \right] + \lambda \cdot R_k \left(f_l \right) \right\} \tag{15}$$

where $R_k(f_l)$ denotes the rate necessary for encoding the filter index l used for prediction in the region k.

3. Scalable quantization of wavelet coefficients

In scalable mesh compression, the wavelet coefficients in the subbands are quantized using a generic family of embedded deadzone scalar quantizers (EDSQ) (Taubman & Marcelin, 2001), in which every wavelet coefficient X is quantized to:

$$q_{\xi,n} = \begin{cases} sign(X) \cdot \left\lfloor \dfrac{|X|}{\Delta_n} + \xi_n \right\rfloor & if \ \dfrac{|X|}{\Delta_n} + \xi_n > 0 \\ 0 & otherwise \end{cases} \tag{16}$$

where $n \in \mathbf{Z}_+$ denotes the quantization level. ξ_n and Δ_n denote the deadzone control parameter and the step size for any $n \geq 0$, respectively, with $\xi_n = \xi_0/2^n$ and $\Delta_n = 2^n \Delta_0$, where ξ_0 and Δ_0 are the parameters for the highest rate quantizer $(n = 0)$. Note that $\xi_0 = 0$ corresponds to the well-known SAQ (Shapiro, 1993) in which the deadzone size is twice the step size Δ_n for any n.

3.1 Wavelet coefficient histogram

In general, the observed histogram H_k^j of the kth, $k \in \{x, y, z\}$, coordinate component of the jth wavelet subband is symmetric around its center of mass which is often zero or very close to zero. Moreover, the histogram is peaky around the mean and the frequency of occurrence decays as the magnitude of the coefficient's component increases. Fig. 5 depicts the observed histograms of the d^{J-3} subband of *Rabbit* (*non-normal* mesh) and *Dino* (*normal* mesh) obtained using the classical Butterfly transform. It is observed experimentally that, in general, $\sigma^2(H_k^{j+1}) < \sigma^2(H_k^j)$ for $1 \leq j < J$.

In the literature, the observed histogram of any component of a wavelet subband is generally modeled using a zero mean generalized Gaussian (GG) distribution (Mallat, 1989), expressed by:

$$\forall x \in \mathbb{R} \quad f_{GG}(x,\sigma,\alpha) = \frac{\alpha \upsilon^{1/\alpha}}{2\Gamma(1/\alpha)} e^{-\upsilon|x|^\alpha} , \tag{17}$$

where α, $\alpha \in (0,2]$, is the shape control parameter. $\upsilon > 0$ is the scaling factor and $\upsilon^{1/\alpha} = \sqrt{\Gamma(3/\alpha)/\sigma^2 \Gamma(1/\alpha)}$, where Γ is the Gamma function. Note that, for $\alpha = 1$, Eq (17) transforms into a zero-mean Laplacian probability density function (PDF) given by:

$$\forall x \in \mathbb{R} \quad f_L(x,\sigma) = \frac{1}{\sigma\sqrt{2}} e^{-\frac{\sqrt{2}}{\sigma}|x|} = \frac{\lambda}{2} e^{-\lambda|x|} \text{ where } \lambda = \frac{\sqrt{2}}{\sigma} , \tag{18}$$

and for $\alpha = 2$ Eq (17) corresponds to a zero-mean Gaussian PDF.

Although GG distributions closely approximate the observed histogram of wavelet coefficients, only approximate rate and distortion expressions for a uniformly quantized GG random variable are known (Fraysse et al,. 2008). The extension of these expressions to embedded quantization is not evident as the rate and distortion functions for such distributions are not easily tractable and can only be computed numerically. Moreover, computing these quantities gets very cumbersome due to the slow numerical integration of expressions involving a GG probability function, especially for $\alpha \ll 1$.

3.2 Proposed Laplacian mixture model

In order to avoid the aforementioned drawbacks of GG distributions, we propose a simple Laplacian mixture (LM) model which not only gives an easy closed-form derivation of the distortion and rate quantities but also better approximates the observed histogram of wavelet coefficients in the majority of cases. The proposed LM is a linear combination of two Laplacian PDFs, i.e.,

$$\forall x \in \mathbb{R} \quad f_{LM}(x) = \beta \cdot f_L(x,\sigma_1) + (1-\beta) \cdot f_L(x,\sigma_2) . \tag{19}$$

Note that $f_{LM}(x)$ indeed defines a probability function, as $\int_{-\infty}^{\infty} f_{LM}(x)dx = 1$.

The LM model is fitted over the observed data using the expectation maximization (EM) algorithm (Dempster et al., 1977) in order to determine the parameters σ_1, σ_2 and β. The E-step in the EM process calculates two responsibility factors

$$r_1(i) = \frac{\beta \cdot f_L(x_i,\sigma_1)}{\beta \cdot f_L(x_i,\sigma_1) + (1-\beta) \cdot f_L(x_i,\sigma_2)}, \quad r_2(i) = \frac{(1-\beta) \cdot f_L(x_i,\sigma_2)}{\beta \cdot f_L(x_i,\sigma_1) + (1-\beta) \cdot f_L(x_i,\sigma_2)},$$

of each observation $x_i, 1 \le i \le N$ and the M-step updates the parameters to be estimated, as:

$$\sigma_m = \sqrt{2} \frac{\sum_{i=1}^{N} r_m(i).|x_i|}{\sum_{i=1}^{N} r_m(i)}, \quad m = 1,2, \quad \text{and} \quad \beta = \frac{1}{N}\sum_{i=1}^{N} r_1(i).$$

The E- and M- steps are executed in tandem till the algorithm achieves minimum *Kullback-Leibler* (KL) distance between the observed and model histograms. A better convergence rate is achieved by the initialization condition $\sigma_1^2 = 0.5\sigma_E^2$, $\sigma_2^2 = 2\sigma_E^2$ and $\beta = 0.9$, where σ_E^2 is the estimated data variance. Histogram fitting for GG distributions is done using the brute-force method where parameters σ_1, σ_2 and β are exhaustively computed for a minimum KL distance.

3.3 Distortion-Rate (D-R) function

Closed-form expressions for the output distortion D_L and the output rate R_L of a Laplacian source quantized using an n level EDSQ are derived in the Appendix. In this section, we derive the D-R function for our proposed LM model. Since the distortion is a linear function of the source PDF, the output distortion D_{LM} of the LM PDF for any quantization level n can be written as:

$$D_{LM}\left(Q_{\delta_n,\Delta_n}\right) = \beta \cdot D_L\left(Q_{\delta_n,\Delta_n}\right) + (1-\beta) \cdot D_L\left(Q_{\delta_n,\Delta_n}\right), \text{ with } \delta_n = 1 - \xi_n. \tag{20}$$

This does not hold for the output rate R_{LM} since the entropy involves the non-linear $\log(.)$ function. Instead, R_{LM} can be computed as an infinite sum:

$$P_0 = 2\int_0^{\delta_n\Delta_n} f_{LM}(x)dx, \ P_k = \int_{(k-1+\delta_n)\Delta_n}^{(k+\delta_n)\Delta_n} f_{LM}(x)dx, \ k = 1,2,3..., \text{ and } R_{LM}(Q_{\delta_n,\Delta_n}) = -\sum_{k=-\infty}^{\infty} P_k \log_2 P_k$$

where P_k denotes the probability mass of the kth quantization cell ($k = 0$ corresponds to the deadzone cell). Since the LM model is symmetric around its mean, $P_k = P_{-k}$. Note that the probability mass function (PMF) can be computed exactly due to the possibility of analytical integration of $f_{LM}(x)$. For the GG distribution, however, only numerical integration is possible.

3.4 Model validation

This section demonstrates that the proposed LM model is able to approximate the observed histogram and the observed D-R function of 3D wavelet coefficients more accurately compared to the commonly utilized GG distributions. For comparison purpose, results for the single Laplacian $f_{LM}(\beta = 0)$ case are also reported.

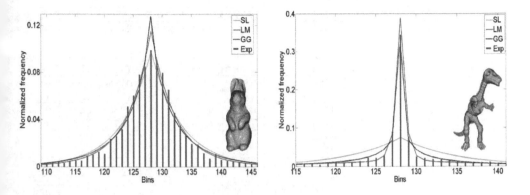

Fig. 5. Probability function fitting over the observed histogram (Exp) for $\mathbf{d}^{l-3}$ -*normal component* for Rabbit (left) and Dino (right). SL is used as the abbreviation of single Laplacian PDF.

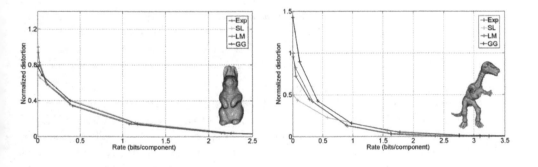

Fig. 6. Modeled and observed D-R functions for the histograms of Fig. 5. Rate is taken as bits per spatial coordinate component.

Fig. 5 illustrates that the proposed mixture model provides a better fitting probability function for the observed histogram compared to the Laplacian and GG distributions. This is especially true for the middle range positive and negative coefficients values – see Fig. 5. For the Rabbit mesh, LM gives only slightly better fitting than the other two models. However, for Dino, the LM can clearly model the fast decay of the observed histogram more accurately than the GG. The Laplacian PDF in this case only gives a very coarse approximation of the observed histogram.

Mesh Type	Mesh (Filter)	SL			LM			GG		
		Nor	Tan 1	Tan 2	Nor	Tan 1	Tan 2	Nor	Tan1	Tan 2
Non-Normal	Venus(U-BF)	0.097 (6.3)	0.114 (10.3)	0.103 (8.9)	0.075 (1.3)	0.100 (3.3)	0.091 (2.8)	0.086 (4.7)	0.102 (5.0)	0.089 (4.7)
	Venus(L-BF)	0.137 (11.4)	0.137 (10.1)	0.104 (6.0)	0.090 (2.0)	0.108 (2.9)	0.080 (1.9)	0.112 (8.1)	0.121 (5.4)	0.092 (4.1)
	Venus(Loop)	0.113 (9.4)	0.102 (7.3)	0.091 (6.7)	0.085 (3.3)	0.090 (1.8)	0.069 (1.7)	0.098 (7.1)	0.092 (3.9)	0.081 (5.0)
	Rabbit(U-BF)	0.170 (8.6)	0.171 (10.4)	0.172 (10.2)	0.134 (1.4)	0.136 (2.0)	0.132 (1.8)	0.150 (5.7)	0.143 (5.2)	0.147 (6.2)
	Rabbit(L-BF)	0.208 (12.0)	0.188 (10.7)	0.177 (8.4)	0.143 (2.5)	0.140 (1.5)	0.138 (1.8)	0.160 (6.7)	0.152 (5.1)	0.153 (5.3)
	Rabbit(Loop)	0.167 (11.2)	0.207 (11.2)	0.173 (8.3)	0.115 (2.4)	0.156 (1.8)	0.135 (2.3)	0.136 (7.9)	0.177 (7.6)	0.152 (5.2)
Normal	Dino(U-BF)	0.527 (16.2)	0.656 (34.3)	0.971 (42.7)	0.145 (5.6)	0.147 (7.8)	0.154 (9.4)	0.165 (7.5)	0.132 (23.4)	0.141 (30.3)
	Skull(U-BF)	1.108 (37.2)	1.473 (44.9)	1.877 (50.4)	0.120 (3.9)	0.138 (7.9)	0.157 (20.5)	0.145 (12.8)	0.141 (15.4)	0.141 (19.9)
	Skrewdriver(U-BF)	0.5294 (33.0)	0.6477 (42.4)	0.6377 (41.9)	0.309 (14.4)	0.251 (17.0)	0.263 (20.0)	0.315 (25.0)	0.262 (34.8)	0.249 (35.2)

Table 1. *KL* (*%ME*, the modeling error as defined in Eq (19)) for the *normal* (*NOR*) and the two *tangential* components (*TAN1*, *TAN2*) averaged over the five subbands. *U-BF* (Unlifted Butterfly), *L-BF* (Lifted Butterfly).

Fig. 6 plots the observed and model D-R curves for the same subband as the one used in Fig. 5. For Rabbit, the LM D-R almost completely overlaps the observed D-R curve. In both cases, the D-R function of the proposed LM model follows the experimental D-R curve more closely than the other two models.

In Table 1, the average *KL* divergence results for the Laplacian, GG and LM models for two non-normal (Venus, Rabbit) and three normal (Dino, Skull, Skredriver) meshes are shown. Each of the three coordinate components is considered separately. For each trial of Table 1, average is taken over five highest resolution subbands. For the large majority of cases, the LM model gives better fitting of the observed histogram than the competing GG model. Note that the Laplacian model gives always the worst fitting results. Also, the LM model gives equally good fitting for both *normal* (*Nor*) and *tangential* (*Tan 1* and *Tan 2*) components. Superior histogram fitting results of our proposed model are also observed for the SAWT of Section 2.3. These results are not reported here due to lack of space.

In Table 1, the percentage modeling error *ME(%)* relative to the *KL* divergence is shown in parenthesis of each table entry. The *ME(%)* is defined in order to gauge the D-R accuracy of the proposed mixture model with respect to other two models. *ME(%)* is defined as:

$$ME(\%) = \frac{\int\limits_{R \in \Re} |D_M(R) - D_E(R)|}{\int\limits_{R \in \Re} \max_R \{D_M(R), D_E(R)\}} \times 100 \ . \tag{21}$$

Mesh Type	Mesh (Filter)	SL			LM			GG		
		$J-1$	$J-2$	$J-3$	$J-1$	$J-2$	$J-3$	$J-1$	$J-2$	$J-3$
Non-Normal	Venus(U-BF)	0.044 (13.6)	0.038 (8.4)	0.039 (3.5)	0.014 (2.9)	0.008 (0.85)	0.025 (1.5)	0.027 (9.5)	0.024 (5.6)	0.033 (2.8)
	Venus(L-BF)	0.050 (13.8)	0.070 (13.9)	0.076 (7.6)	0.010 (2.1)	0.010 (1.5)	0.027 (1.7)	0.029 (9.9)	0.041 (10.1)	0.049 (4.1)
	Venus(Loop)	0.051 (13.8)	0.054 (11.0)	0.038 (4.0)	0.016 (2.4)	0.009 (1.6)	0.023 (0.80)	0.036 (11.0)	0.031 (6.8)	0.032 (2.7)
	Rabbit(U-BF)	0.064 (14.0)	0.062 (11.4)	0.082 (8.1)	0.008 (1.5)	0.011 (1.0)	0.035 (1.3)	0.029 (8.6)	0.032 (6.8)	0.054 (4.9)
	Rabbit(L-BF)	0.069 (14.2)	0.093 (13.9)	0.111 (11.2)	0.007 (1.5)	0.011 (1.1)	0.035 (1.8)	0.029 (8.6)	0.040 (8.6)	0.058 (5.9)
	Rabbit(Loop)	0.082 (16.4)	0.088 (14.7)	0.085 (9.0)	0.011 (2.0)	0.013 (1.9)	0.034 (2.0)	0.042 (11.5)	0.038 (8.7)	0.058 (5.4)
Normal	Dino(U-BF)	1.208 (56.7)	0.873 (46.7)	0.623 (34.7)	0.029 (13.7)	0.074 (12.8)	0.049 (4.6)	0.031 (41.3)	0.042 (32.9)	0.058 (20.4)
	Skull(U-BF)	2.039 (65.7)	1.981 (49.9)	1.832 (32.4)	0.054 (34.2)	0.040 (7.6)	0.076 (5.1)	0.036 (40.2)	0.068 (15.1)	0.066 (8.3)
	Skrewdriver(U-BF)	0.536 (67.0)	0.696 (61.6)	0.483 (39.7)	0.067 (53.2)	0.074 (18.2)	0.065 (7.3)	0.035 (64.0)	0.064 (54.3)	0.101 (25.4)

Table 2. KL (%ME) for three resolution subbands averaged over the three coordinate components.

From Table 1, it is evident that on average the proposed LM model performs better than the GG and Laplacian models also in the ME sense. Better ME results are also obtained for SAWT (not reported here). Hence, the proposed LM model along with the derived D-R function is a better choice for modeling both the histogram and the D-R curve of mesh wavelet coefficients compared to the contemporary models. One notices that, a best histogram fitting in KL sense may not always yield the lowest ME.

Table 2 reports the model validation results for different resolution subbands. For each trial the average is taken across the three spatial coordinate components. It is observed that the GG model performs slightly better for the low-resolution subbands of some meshes. The observed histograms in such cases are more Gaussian-alike, i.e., they have a round top. In general, the LM model faces difficulty in approximating such a round-top histogram due to the peaky nature of each of its Laplacian components; the GG fits well such histograms, as it corresponds to a Gaussian distribution for $\alpha = 2$. Nevertheless, the results show that, on average, the LM model outperforms the Laplacian and the GG models in KL as well as in ME sense.

3.5 Optimal embedded quantization

In this section, conclusions regarding the optimal EDSQ to be used in scalable wavelet-based coding of meshes are drawn. Let z denote the ratio between the deadzone size for $n = 0$ (see Eq. (16)) and the step size for $n \geq 0$ of a general EDSQ. The total average signal-to-noise ratio (SNR) difference which is utilized to measure the performance gap of different embedded quantizers is defined as:

$$\Delta\overline{SNR} = \frac{1}{N}\sum_{\Re}\left(\underset{z=1}{SNR(R)} - \underset{z>1}{SNR(R)}\right),$$

which is computed over a rate range $\Re$ for N rate points, where $SNR(R)$ denotes the discrete SNR-rate function. The $SNR = 10\log_{10}(\sigma^2/D)$ is computed in dBs, where D is the total distortion in the transform domain. The difference in SNR is computed relative to the uniform embedded quantizer (UEQ), i.e., $z=1$. $\Delta\overline{SNR}$ for five embedded deadzone quantizers is plotted in Fig. 7. over a wide range of standard deviation ratios σ_2/σ_1. In Fig. 7., the commonly observed proportion $\beta = 0.9$ is considered, as mentioned in Section 3.2.

We determined experimentally that at lower standard deviation ratios, $\Delta\overline{SNR}$ is positive and the UEQ is optimal for $\sigma_2/\sigma_1 < 120$. For $120 < \sigma_2/\sigma_1 < 290$, the quantizer with $z = 1.5$ performs better compared to all other quantizers. Similarly, $z = 2$ (i.e. the SAQ) performs the best in the range $290 < \sigma_2/\sigma_1 < 600$, while $z = 2.5$ performs the best for $600 < \sigma_2/\sigma_1$. In general, small standard deviation ratios correspond to α close to 1, observed in non-normal meshes, while higher ratios correspond to $\alpha \ll 1$, observed in normal meshes. These results show that one cannot determine a single embedded quantizer that provides the best performance for *all* 3D meshes. However, an optimal quantizer per wavelet coordinate can be determined based on the corresponding σ_2/σ_1 extracted from the model.

Overall, for $\sigma_2/\sigma_1 < 120$, the difference between SAQ and the UEQ is significant, and hence UEQ is the optimal choice. For $\sigma_2/\sigma_1 \geq 120$, SAQ is not always the optimum, but lies not far from the optimum.

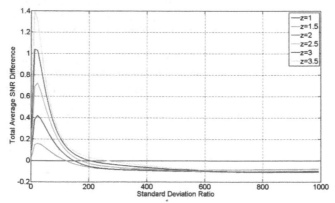

Fig. 7. SNR difference for five EDSQs with respect to UEQ.

Given the fact that SAQ is closely linked to bit-plane coding and that it can be implemented using simple binary arithmetic, one concludes that SAQ is not an optimal, but an acceptable solution in scalable coding of meshes.

4. Analysis of wavelet coefficient dependencies

Similar to images, parent-children and neighboring wavelet coefficient dependencies exist in wavelet decomposed mesh structure. In Fig. 8 (middle, right), the positions of the wavelet coefficients at different levels of the transform are shown with the help of white and dark circles. In particular, wavelet coefficients have a one-to-one correspondence with the edges of the coarser mesh. For each wavelet coefficient there are rings of neighboring coefficients which lie in the same wavelet subband – see Fig. 8 (right). Also, a set of four wavelet coefficients have a parent coefficient at the next coarser resolution – see Fig. 8 (middle, right).

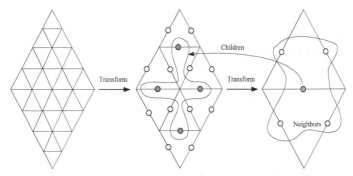

Fig. 8. Parent-children and neighboring wavelet coefficients: actual mesh (left); coarser meshes after one (middle), and after two wavelet decomposition levels (right).

Statistical intraband dependencies exist between neighboring coefficients of each resolution level. The main reason for the existence of these dependencies is the smoothness of the surface. Wavelet coding paradigms that exploit the intraband dependencies between the wavelet coefficients are known as intraband wavelet codecs such as block-based coding techniques (Munteanu et al., 1999a), quadtree coding approaches (Munteanu et al., 1999b), and the EBCOT codec employed in the JPEG-2000 scalable image coding standard (Taubman, 2000).

Statistical dependencies also exist between the parent and descendants (children) due to the natural decay of the coefficients' magnitude for increasing frequencies. In other words, if a parent coefficient magnitude is below a certain threshold, then there is a high probability that the magnitude of its descendants will be also below this threshold. This corresponds to the so-called zerotree-model, firstly introduced by Shapiro in (Shapiro, 1993). The wavelet coding paradigms that exploit the parent-children dependencies are known as interband wavelet codecs.

Finally, there is a third category of coding paradigms, exploiting both the interband and intraband statistical dependencies between the wavelet coefficients. They are generally known as composite codecs, EZBC (Hsiang & Woods, 2000) and the ECECOW approach of (Wu, 1997) are typical examples of codecs in this category.

In the following, an information theoretical analysis of the aforementioned coefficient dependencies is presented. Our aim is to single out the type of dependencies which can ensure best compression performance in the context of wavelet-based mesh compression.

4.1 Mutual information analysis

The mutual information is the reduction in the entropy of one random variable due to the knowledge of the other random variable.

$$I(X;Y) = h(X) - h(X/Y) ,$$
(22)

It is known that $I(X,Y) = I(Y,X)$. In the wavelet domain, we define the following mutual information quantities:

$I(X;P_X)$: denotes the mutual information between a wavelet coefficient X and its parent coefficient P_X.

$I(X;\mathbf{n}_X)$: denotes the mutual information between a wavelet coefficient X and its neighboring wavelet coefficients $\mathbf{n}_X = [n_{1,X}, n_{2,X}, \dots n_{N,X}]$.

$I(X;P_X;\mathbf{n}_X)$: denotes the composite mutual information.

From the basics of information theory (Cover & Thomas, 1991), we know that:

$$I(X;P_X;\mathbf{n}_X) \geq I(X;\mathbf{n}_X) \quad \text{and} \quad I(X;P_X;\mathbf{n}_X) \geq I(X;P_X).$$
(23)

For the estimation of $I(X;\mathbf{n}_X)$, we need to estimate the joint PDF $p(x,\mathbf{n}_x)$ which can have high dimensionality depending on the number of considered neighbors. Since the amount of data needed to accurately estimate a PDF increases exponentially with its dimensionality, it is difficult to reliably estimate a high-dimensional PDF. To alleviate this problem, the reduction in dimensionality as proposed in (Liu & Moulin, 2000) is used here. We summarize the neighborhood of X through a so-called summarizing function $T = g(\mathbf{n}_X)$. This function maps the neighboring wavelet coefficients to a single value. We note that such a many-to-one summarizing function cannot increase the mutual information, i.e.,

$$I(X;\mathbf{n}_X) \geq I(X;T) .$$
(24)

Equality in the above equation holds if $\mathbf{n}_X \to T \to X$ forms a Markov chain. The summarizing function used in our analysis is:

$$T = f(\mathbf{n}_X) = \sum_{i=1}^{N} (n_{i,X})^2 .$$
(25)

Due to this summarizing function, it is sufficient to compute the joint PDF $p(x,t)$, t is a realization of the random variable T, instead of $p(x,\mathbf{n}_x)$, for the estimation of the intraband mutual information $I(X;\mathbf{n}_X)$.

In our analysis, the mutual information for the defined quantities is estimated using the adaptive partitioning method (Darbellay & Vajda, 1999) instead of the traditional histogram method. This is because the histogram method highly depends on the bin size and for a small bin size there may not be sufficient number of observations in some bins to make a correct estimate. The adaptive partitioning method (Darbellay & Vajda, 1999) on the other hand, ensures that there are always sufficient numbers of observations in each bin, and provides reliable estimates of the mutual information.

Mesh Type		Butterfly			Loop		
	MESH	INTRABAND	INTERBAND	COMPOSITE	INTRABAND	INTERBAND	COMPOSITE
Non-Normal	Venus	0.3727	0.1902	0.6886	0.8320	0.5591	1.5847
	Bunny	0.3960	0.1992	0.6844	0.8033	0.5628	1.5427
	Horse	0.5615	0.2869	0.9873	1.0482	0.6943	1.9684
	Rabbit	0.4048	0.2017	0.7089	0.8996	0.6450	1.7425
	Feline	0.8277	0.2134	1.0696	1.1471	0.6285	2.0287
Normal	Venus	0.3052	0.2130	0.5741	-	-	-
	Skull	0.3381	0.2922	0.7001	-	-	-
	Dino	0.3043	0.2804	0.6672	-	-	-

Table 3. Average mutual information in bits for several non-normal and normal meshes.

Table 3 shows the average mutual information results for interband, intraband and composite dependencies for various mesh models. Since in mesh coding three different components need to be coded for each vertex position in space, the average mutual information $I_{avg} = (I_X + I_Y + I_Z)/3$ is reported instead of the mutual information for the three components individually. It is observed from Table 3 that for both normal and non-normal meshes mutual information of interband models is the least, independent of the wavelet transform employed. On the other hand mutual information for intraband models is significantly higher than for the interband models. Finally, composite models which gather the characteristics of both interband and intraband models exhibit even higher mutual information than interband or intraband models alone. Mathematically we can summarize our numerical findings as:

$$I(X;P_X) \ll I(X;\mathbf{n}_X) \ll I(X;P_X;\mathbf{n}_X). \tag{26}$$

Experimental results for the mutual information based estimation of interband, intraband and composite dependencies seem to indicate that exploiting the composite dependencies should be preferred. Additionally, it is important to point out that favoring intraband over zerotree-based interband models brings along the additional benefit of resolution scalability. Specifically, by following an intraband codec design, only those wavelet subbands that are needed in order to reconstruct a target mesh resolution-level need to be encoded, while the others can be discarded. This does not hold in case of interband and composite codec designs, due to the tree-structures that span all the wavelet decomposition levels. Since composite models cannot be discarded altogether due to their highest mutual information property, a careful implementation of a composite mesh coding system needs to be carried out into order to get the benefit of both the higher compression efficiency and the resolution scalable decoding at the same time.

Finally, it is important to point out that the differences in terms of mutual information do not give any indication about the final performance differences between interband, intraband and composite coding systems. Hence, an actual development and comparison of such coding systems is needed in order to experimentally validate the conclusions of this information-theoretic analysis of wavelet-based mesh coding designs, which is presented next.

5. Scalable mesh compression overview

In this section, we give a brief overview of the scalable mesh compression systems. Based on the design choices established earlier, we designed intraband and composite mesh coding systems which provide state-of-the-art compression performance, together with resolution as well as quality scalability of the compressed mesh.

5.1 Progressive Geometry Compression (PGC)

The first scalable wavelet-based geometry compression technique is the progressive geometry compression (PGC) codec proposed by Khodakovsky et al. in (Khodakovsky et al., 2000). PGC makes use of the well-know zero-tree coding (Shapiro, 1993) of wavelet coefficient's bitplanes in order to encode the decomposed mesh structure. Significant improvements in the compression performance against the contemporary scalable as well as non-scalable mesh coding systems were reported in (Khodakovsky et al., 2000). However, a major drawback of PGC schemes is their inability to provide resolution scalability. This is caused by the zero-tree structure which, for a given bitplane, spans all the wavelet decomposition levels. For a detailed understanding of the PGC system we refer to (Khodakovsky et al., 2000).

5.2 Scalable Intraband Mesh Compresion (SIM)

Despite of the great success of zerotree-based coding techniques in image coding, the choice of an interband codec design is not necessarily the best option in the context of scalable mesh coding. This was illustrated in Section 4 where different types of dependencies among wavelet coefficients were studied. Based on this analysis, we opt for an intraband dependency model in our codec design. As mentioned before, favoring intraband models over interband models brings along the additional benefit of resolution scalability. Specifically, by following an intraband codec design, only those wavelet subbands that are needed in order to reconstruct a target mesh resolution-level need to be encoded, while the others can be discarded.

In the designed scalable intraband mesh (SIM) compression system (Denis et al., 2010b) each resolution subband is encoded independently of the others. Similar to (Shapiro, 1993), SAQ is applied to each resolution subband to determine the significance of the wavelet coefficients with respect to a series of monotonically decreasing thresholds. Based on the significance outcome, a tree node is split into eight equal volume nodes. The resulting octree nodes may contain an unequal number of wavelet coefficients. In general, the number of coefficients in all nodes of a same tree-depth is roughly the same. This way, an octree is constructed for each resolution subband, wherein the depth of the tree (number of levels in the octree) is equal to the number of bitplanes of the subband. All magnitude bitplanes are sequentially coded using the non-significance, the significance and the refinement coding passes. For a detailed presentation of the SIM codec the interested reader is referred to (Denis et al., 2010b).

Using the octree-based bitplane coding, separate symbol streams are first generated for all bitplanes of each resolution subband. Depending on the type of scalability, i.e., resolution or quality scalability, the encoded symbol streams are entropy coded using a predefined progression order of bitplanes. For quality scalability, bitplanes of certain significance, from all resolution subbands, are first encoded before encoding the bitplanes of lower significance. However, in resolution scalability mode, all bitplanes of a lower resolution subband are progressively encoded before encoding the next higher resolution subband.

We compared the SIM codec with the PGC codec for both normal and non-normal 3D meshes. The decoded meshes are compared against the original semi-regular input meshes using the peak signal-to-noise ratio (PSNR) as the distortion metric, which is defined as:

$$PSNR = 20 \cdot \log_{10}\left(\frac{peak}{RMS}\right)(dBs),$$

where *peak* and *RMS* denote the size of the bounding box and the root mean squared error calculated on the distances between the decoded vertex positions with respect to the original ones, respectively.

Fig. 9 depicts PSNR versus bitrate (bits per semi-regular vertex) plots, evaluated for the semi-regular non-normal Venus and Bunny meshes using the Butterly transform. The results demonstrate that for both meshes, SIM yields superior performance when compared to PGC.

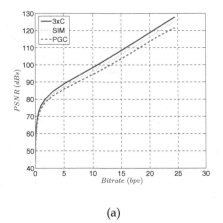
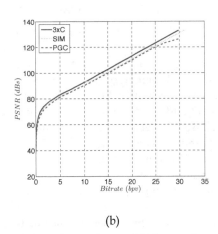

(a)   (b)

Fig. 9. PSNR versus bitrate for non-normal mesh models in the quality scalability mode: (a) Venus, (b) Bunny . The lifted Butterly transform is employed for all three codecs.

The averaged gain in PSNR when compressing the Venus and Bunny meshes goes up to 2.22 dB and 2.35 dB, respectively. One may also notice the increasing performance difference with increasing bitrates; this indicates that the SIM coder tends to code the high frequency information more efficiently. For the spatially adaptive wavelet transform (SAWT) the compression results are reported in (Denis et al., 2010a).

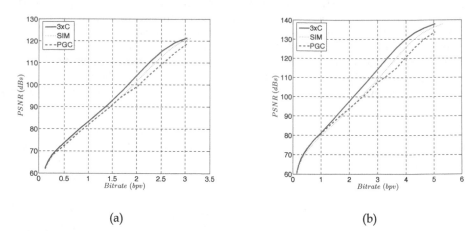

(a)   (b)

Fig. 10. PSNR versus bitrate for normal mesh models in the quality scalability mode: (a) Skull, (b) Dino. The un-lifted Butterly transform is employed for all three codecs.

Fig. 10 shows compression performance plots for two normal meshes, Skull and Dino. One notices that at low bitrates, PGC tends to compress better. However, the ability of SIM to capture and code more efficiently the high-frequency components is noticeable at high bitrates and leads to an improved performance when compared to PGC.

5.3 Composite Context-conditioned Compression (3xC)

The mutual information analysis presented earlier showed that the composite dependencies between the wavelet coefficients are by far the strongest. However, one may notice that, employing composite models may hinder, similar to interband models, the possibility of providing resolution scalability. Thus one must be careful in exploiting the parent-children dependencies within composite models. A careful observation reveals that exploiting parent-children dependencies in a causal fashion (Denis et al., 2010b) does not limit resolution scalable decoding of the compressed mesh. Following this observation, we proposed a scalable composite mesh compression system in (Denis et al., 2009), (Denis et al., 2010b). The bitplane coding modules of the SIM codec and the 3xC codec are identical. The two designs differ at the entropy coding level. In particular, for 3xC, parent coefficient based context-conditioning is employed in the entropy coding module. For context-conditioning, significant, non-significant as well as sign information is entropy coded using the designed context tables. The refinement information is encoded without context-conditioning; this is because including the parental information when entropy coding the refinement symbols does not improve compression performance. For a detailed presentation of the 3xC codec the interested reader is referred to (Denis et al., 2009).

Fig. 9 also depicts the PSNR curves computed for the non-normal Venus and Bunny meshes using our implementation of the un-lifted butterfly based 3xC mesh compression system. The figure clearly demonstrates that, when dealing with non-normal meshes, 3xC systematically yields superior performance compared to PGC as well as SIM.

In the case of normal meshes (Fig. 10) our coder employs the same transform as PGC. Both codecs perform the same at very low bitrates. However, overall, 3xC yields the best compression performance. 3xC gives approximately equivalent results when compared with the intraband SIM codec for normal meshes. This is because the context-conditioning is only possible for the normal component of vector valued wavelet coefficients. Overall, it is clear that the proposed 3xC codec produces similar, and in almost all cases, superior performance compared to PGC and SIM codecs.

5.4 Visual comparison: PGC vs 3xC

Visual comparisons of Bunny and Skull meshes, compressed and reconstructed using 3xC at different bits per vertex (bpv), are presented in Fig. 11 and Fig. 12, respectively. The colored regions highlight the distortions introduced by lossy compression. For low-to-medium bitrates, the pure red color indicates areas where the distance between the original and decoded vertex is larger than 0.1% of the diagonal of the bounding box of the semi-regular mesh. For high bitrates, the distortion is visualized with respect to 0.02% of the diagonal. The mesh is shaded greener as the distortion lowers, with pure green indicating no distortion.

When visually comparing the compressed Bunny and Skull meshes produced by 3xC and PGC, it is very clear that 3xC yields superior performance for all bitrates. Taking the result at 0.050 bpv as an example, we observe that many areas which are shaded red for PGC are green for 3xC. At high rates, the differences between the mesh geometries may not be visually significant, yet the colors reveal that 3xC is able to approximate the original mesh much more accurately compared to the PGC system.

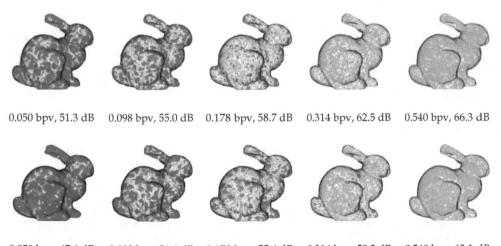

0.050 bpv, 51.3 dB   0.098 bpv, 55.0 dB   0.178 bpv, 58.7 dB   0.314 bpv, 62.5 dB   0.540 bpv, 66.3 dB

0.050 bpv, 47.6 dB   0.098 bpv, 51.4 dB   0.178 bpv, 55.4 dB   0.314 bpv, 59.2 dB   0.540 bpv, 63.1 dB

Fig. 11. Visual comparison of non-normal Bunny mesh using (top row) the 3xC codec and (bottom row) the PGC codec. The red color intensity reflects the distortion with respect to the uncompressed semi-regular mesh. The rate for the base mesh (i.e., M^0 - see section 2.1.2) is not included in the reported rate values.

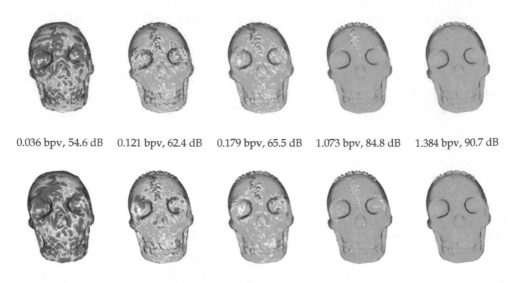

0.036 bpv, 54.6 dB   0.121 bpv, 62.4 dB   0.179 bpv, 65.5 dB   1.073 bpv, 84.8 dB   1.384 bpv, 90.7 dB

0.036 bpv, 53.9 dB   0.121 bpv, 61.9 dB   0.179 bpv, 65.2 dB   1.073 bpv, 83.4 dB   1.384 bpv, 88.9 dB

Fig. 12. Visual comparison of normal Skull mesh using (top row) the 3xC codec and (bottom row) the PGC codec. The red color intensity reflects the distortion with respect to the uncompressed semi-regular mesh. The rate for the base mesh is not included in the reported rate values.

The visual comparisons of the normal mesh Skull at different bpv are shown in Fig. 12. Though, at first glance it may appear that both codecs perform very similar, small differences are noticeable when investigating the meshes more closely. When examining the comparison at 0.036 bpv, we notice that the PGC codec preserves more details in Skull's teeth. The green shade for 3xC at rate 0.179 bpv, however, seems more pure compared to PGC for which it is rather yellowish green. We also observe that no red regions are present for 3xC at rate 1.073 bpv, whereas some are visible for PGC at the same rate.

6. Conclusions

In this book chapter, we propose a constructive methodology for the design of scalable wavelet-based mesh compression systems. Our design strategy differs from conventional designs which simply opt for reusing methods from wavelet-based image coding for the design of mesh coding systems. In particular, our methods are motivated by an information-theoretic analysis of the statistical dependencies between wavelet coefficients which shows that, intraband dependencies are systematically stronger than interband ones for both normal and non-normal meshes, and that composite models are the best. We also investigate the optimality of successive approximation quantization, commonly used in scalable compression, in the context of wavelet-based mesh compression. Using a Laplacian mixture model, it is shown that successive approximation quantization is an acceptable, but in general not an optimal solution. Anchored in these results, novel intraband and composite coding systems are presented which improve the state-of-the-art in scalable mesh compression, both in terms of scalability and compression efficiency.

7. Appendix

The output distortion D_L of a Laplacian PDF, quantized using an n level EDSQ and reconstructed using midpoint reconstruction, can be written as:

$$D_L(Q_{\xi_n,\Delta_n}) = \lambda \underbrace{\int_0^{(1-\xi_n)\Delta_n} x^2 e^{-\lambda x}\, dx}_{D_{DZ}} + \lambda \underbrace{\sum_{k=1}^{\infty} \int_{(k-\xi_n)\Delta_n}^{(k+1-\xi_n)\Delta_n} \left(x-(k+0.5-\xi_n)\Delta_n\right)^2 e^{-\lambda x}\, dx}_{D_{REST}},$$

where D_{DZ} and D_{REST} denote the distortion contributions of the deadzone and the other quantization cells, respectively. By proper substitution and letting

$$\sum_{k=1}^{\infty} e^{-\lambda\Delta_n k} = \frac{e^{-\lambda\Delta_n}}{1-e^{-\lambda\Delta_n}}, \text{ as } e^{-\lambda\Delta_n} \leq 1, \tag{27}$$

the following closed-form expression for the distortion is obtained:

$$D_L(Q_{\delta_n,\Delta_n}) = \frac{2}{\lambda^2} + e^{-\lambda\Delta_n\delta_n}\left\{ \left(\frac{1}{4}-\delta_n^2\right)\Delta_n^2 - \left(2\delta_n + \coth\left(\frac{\lambda\Delta_n}{2}\right)\right)\frac{\Delta_n}{\lambda}\right\}, \tag{28}$$

where $\delta_n = 1-\xi_n$.

Similarly, the output rate R_L of a Laplacian PDF, quantized using an n level EDSQ can be written as:

$$R_L(Q_{\xi_n,\Delta_n}) = -2\underbrace{\left(\frac{\lambda}{2}\int_0^{(1-\xi_n)\Delta_n} e^{-\lambda x}\, dx\right)\log_2 2\left(\frac{\lambda}{2}\int_0^{(1-\xi_n)\Delta_n} e^{-\lambda x}\, dx\right)}_{R_{DZ}} \cdots$$

$$\cdots -2\underbrace{\sum_{k=1}^{\infty}\left(\frac{\lambda}{2}\int_{(k-\xi_n)\Delta_n}^{(k+1-\xi_n)\Delta_n} e^{-\lambda x}\, dx\right)\log_2\left(\frac{\lambda}{2}\int_{(k-\xi_n)\Delta_n}^{(k+1-\xi_n)\Delta_n} e^{-\lambda x}\, dx\right)}_{R_{REST}}.$$

Again making use of the summation reduction identity of (27) along with the identity

$$\sum_{k=1}^{\infty} e^{-\lambda\Delta_n k}\log_2\left(e^{-\lambda\Delta_n k}\right) = \log_2\left(e^{-\lambda\Delta_n}\right)\sum_{k=1}^{\infty} k\left(e^{-\lambda\Delta_n}\right)^k = \frac{\log_2\left(e^{-\lambda\Delta_n}\right)e^{-\lambda\Delta_n}}{\left(1-e^{-\lambda\Delta_n}\right)^2},$$

the expression for the rate can be reduced to the following closed-form:

$$R_L(Q_{\delta_n,\Delta_n}) = c_\delta \log_2\left(\frac{2d_{\delta_n}}{d_1 c_1^{1/d_1} e^{\lambda\Delta_n(1-\delta_n)}d_{\delta_n}^{1/c_{\delta_n}}}\right), \tag{29}$$

where $c_{\delta_n} = e^{-\lambda\Delta_n\delta_n}$ (hence $c_1 = e^{-\lambda\Delta_n}$) and $d_{\delta_n} = 1-c_{\delta_n}$ (hence $d_1 = 1-c_1$).

8. Acknowledgements

The authors would like to thank Cyberware, Headus, The Scripps Research Institute, Washington University, and Stanford University for providing 3D models. The authors are particularly grateful to Igor Guskov for providing them with the normal meshes, and to Andrei Khodakovsky and Peter Schröder for providing the PGC software.

This research was supported by the Agency for Innovation by Science and Technology (IWT) - Flanders (OptiMMa project) and the Fund for Scientific Research – Flanders (postdoctoral mandate Peter Schelkens and project G014610N).

9. References

Catmull, E. & Clark, J. (1978). Recursively Generated B-Spline Surfaces on Arbitrary Topological Surfaces. *Computer-Aided Design*. Vol. 10, No. 6, (November 1978), pp. 350-355, ISBN 1-58113-052-X.

Chang, C.-L. & Girod, B. (2006). Direction-Adaptive Discrete Wavelet Transform Via Directional Lifting and Bandeletization. Proc. *IEEE International Conference on Image Processing*, pp. 1149-1152, Atlanta, GA, USA.

Cover, T. M. & Thomas, J. A. (1991). *Elements of Information Theory*. Wiley-Interscience, ISBN 0-471-24195-4, New York, USA.

Darbellay, G. A. & Vajda, I. (1999). Estimation of the Information by an Adaptive Partitioning of the Observation Space. *IEEE Transactions on Information Theory*. Vol. 45, No. 4, (May 1999), pp. 1315-1321, ISSN 0018-9448.

Dempster, A. P., Laird, N. M. & Rubin, D. B. (1977). Maximum Likelihood from Incomplete Data Via the Em Algorithm. *Journal of The Royal Statistical Society, Series B*. Vol. 39, No. 1, (May 1977), pp. 1-38.

Denis, L., Ruxandra, F., Munteanu, A. & Schelkens, P. (2010a). Spatially Adaptive Bases in Wavelet-Based Coding of Semi-Regular Meshes. *Proceedings of the SPIE*, pp. 772310-772310-8, Brussels, Belgium.

Denis, L., Satti, S. M., Munteanu, A., Cornelis, J. & Schekens, P. (2010b). Scalable Intraband and Composite Coding of Semi-Regular Meshes. *IEEE Transactions on Multimedia*. Vol. 12, No. 8, (December 2010), pp. 773-789, ISSN 1520-9210.

Denis, L., Satti, S. M., Munteanu, A., Cornelis, J. & Schelkens, P. (2009). Context-Conditioned Composite Coding of 3D Meshes Based on Wavelets on Surfaces. *IEEE International Conference on Image Processing*, pp. 3509-3512, ISSN 1522-4880, Cairo, Egypt, November 2009.

Dyn, N., Levin, D. & Gregory, J. A. (1990). A Butterfly Subdivision Scheme for Surface Interpolation with Tension Control. *ACM Transactions on Graphics*. Vol. 9, No. 2, (April 1990), pp. 160-169, ISSN 0730-0301.

Eck, M., Derose, T., Duchamp, T., Hoppe, H., Lounsbery, M. & Stuetzle, W. (1995). Multiresolution Analysis of Arbitrary Meshes. *ACM SIGGRAPH, Proceedings of the 22rd Annual Conference on Computer Graphics and Interactive Techniques*, pp. 173-182 ISBN 0-89791-701-4, Los Angeles, California, USA, August 6-11, 1995.

Eck, M. & Hoppe, H. (1996). Automatic Reconstruction of B-Spline Surfaces of Arbitrary Topological Type. *ACM SIGGRAPH, Proceedings of the 23rd Annual Conference on Computer Graphics and Interactive Techniques*, pp. 325–334, ISBN 0-89791-746-4.

Fraysse, A., Pesquet-Popescu, J. C. (2008) Rate-distortion Results on Generalized Gaussian Distributions, *IEEE International Conference on Acoustic, Speech Signal Processing*, pp. 3753-3756, Las Vegas, NV, USA, March 30-April 04, 2008.

Hsiang, S.-T. & Woods, J. W. (2000). Embedded Image Coding Using Zeroblocks of Subband/Wavelet Coefficients and Context Modeling. *IEEE International Symposium on Circuits and Systems*, pp. 662-665, Geneva, CH, May 28-31.

Khodakovsky, A., Schröder, P. & Sweldens, W. (2000). Progressive Geometry Compression. *ACM SIGGRAPH, 27th International Conference on Computer Graphics and Interactive Techniques*, pp. 271-278, ISBN 1-58113-208-5, New Orleans, Lousiana, USA, July 23-28, 2000.

Li, J. & Kuo, C. C. J. (1998). Progressive Coding of 3-D Graphic Models. *Proceedings of the IEEE* Vol. 86, No. 6, pp. 1052-1063, ISSN 0018-9219.

Liu, J. & Moulin, P. (2000). Analysis of Interscale and Intrascale Dependencies between Image Wavelet Coefficients. *International Conference on Image Processing*, pp. 669-671, Vancouver, Canada, September 11-13, 2000.

Loop, C., Schaefer, S., Ni, T. & Castaño, I. (2009). Approximating Subdivision Surfaces with Gregory Patches for Hardware Tessellation. *ACM Transactions on Graphic, SIGGRAPH Asia*. Vol. 28, No. 5, (December 2009), pp. 151:1 - 151:9, ISSN 0730-0301.

Lounsbery, M., Derose, T. D. & Warren, J. (1997). Multiresolution Analysis for Surfaces of Arbitrary Topological Type. *ACM Transactions on Graphics*. Vol. 16, No. 1, (January 1997), pp. 34-73.

Mallat, S. G. (1989). A Theory for Multiresolution Signal Decomposition: The Wavelet Representation. *IEEE Transactions on Pattern Analysis and Machine Intelligence*. Vol. 11, No. 7, (July 1989), pp. 674-693, ISSN 0162-8828.

Munteanu, A., Cornelis, J. & Cristea, P. (1999a). Wavelet-Based Lossless Compression of Coronary Angiographic Images. *IEEE Transactions on Medical Imaging*. Vol. 18, No. 3, (March 1999), pp. 272-281.

Munteanu, A., Cornelis, J., Van der Auwera, G. & Cristea, P. (1999b). Wavelet Image Compression - the Quadtree Coding Approach. *IEEE Transactions on Information Technology in Biomedicine*. Vol. 3, No. 3, (September 1999), pp. 176-185, ISSN 1089-7771.

Pajarola, R. & Rossignac, J. (2000). Compressed Progressive Meshes. *IEEE Transactions on Visualization and Computer Graphics*. Vol. 6, No. 1, (January-March 2000), pp. 79-93, ISSN 1077-2626.

Ronfard, R. & Rossignac, J. (1996). Full-Range Approximation of Triangulated Polyhedra. *Proceeding of Eurographics, Computer Graphics Forum*. Vol. 15, No. 3, (August 1996), pp. 67-76, ISSN 0167-7055.

Rossignac, J. & Borrel, P. (1993). *Multi-Resolution 3d Approximation for Rendering Complex Scenes*. Springer-Verlag, ISBN 0387565299.

Schröder, P. & Sweldens, W. (1995). Spherical Wavelets: Efficiently Representing Functions on the Sphere. *ACM SIGGRAPH, Proceedings of 22nd Annual Conference on Computer Graphics and Interactive Techniques*, pp. 161-172, ISBN 0-89791-701-4, Los Angeles, California, USA, August 6-11.

Shapiro, J. M. (1993). Embedded Image Coding Using Zerotrees of Wavelet Coefficients. *IEEE Transactions on Signal Processing*. Vol. 41, No. 12, (1993), pp. 3445-3462.

Soucy, M. & Laurendeau, D. (1996). Multiresolution Surface Modeling Based on Hierarchical Triangulation. *Computer Vision and Image Understanding*. Vol. 63, No. 1, (January 1996), pp. 1-14, ISSN 10773142.

Taubin, G., Guéziec, A., Horn, W., William, A. & Lazarus, F. (1998). Progressive Forest Split Compression. *ACM SIGGRAPH, Proceedings of 25th International Conference on Computer Graphics and Interactive Techniques*, pp. 123-132, ISBN 0-89791-999-8, Orlando, Florida, USA, July 19-24, 1998.

Taubman, D. (2000). High Performance Scalable Image Compression with EBCOT. *IEEE Transactions on Image Processing*. Vol. 9, No. 7, (July 2000), pp. 1158-1170, ISSN 1057-7149.

Taubman, D. & Marcelin, M. (2001). *Jpeg2000: Image Compression Fundamentals, Standards and Practice*. Springer, ISBN 978-0792375197.

Wu, X. (1997). High-Order Context Modeling and Embedded Conditional Entropy Coding of Wavelet Coefficients for Image Compression. *31st Asilomar Conference on Signals, Systems and Computers*, pp. 1378-1382, Pacific Grove, CA, November 2-5, 1997.

9

Robust Lossless Data Hiding by Feature-Based Bit Embedding Algorithm

Ching-Yu Yang

Department of Computer Science and Information Engineering
National Penghu University of Science and Technology
Taiwan

1. Introduction

Recently, data hiding, or information hiding, plays an important role in data assurance. Generally speaking, data hiding techniques can be classified into steganography and digital watermarking (Cox et al., 2008; Shih, 2008). The marked images generated by the steganographic methods (Gu & Gao, 2009; Liu & Shih, 2008; Qu et al., 2010; Wang et al., 2010; Zhou et al., 2010; Fan et al., 2011) were prone to catch damage (by manipulations) and resulted in a failure extraction of the message. However, based on the spatial domain, the steganographic methods often provide a large payload with a good perceived quality. Major applications of the techniques can be found in private data saving, image tagging and authentication, and covert communications. On the other hand, the robustness performance with a limited payload is a key feature of digital watermarking approaches (Lai et al., 2009; Al-Qaheri et al., 2010; Lin & Shiu, 2010; Yamamoto & Iwakiri, 2010; Yang et al., 2010; Martinez-Noriega et al., 2011). Most of the robust watermarking approaches which based on the transform domain such as discrete cosine transform (DCT), integer wavelet transform (IWT), and discrete Fourier transform (DFT) can be tolerant of common image processing operations. Their usages can be found in owner identification, proof of ownership, and copy control. Note that conventional data hiding techniques were irreversible, namely, the host media can not be recovered after data extraction. To preserve or protect the originality of the valuable (or priceless) host media, for example, military or medical images, and law enforcement, the reversible data hiding schemes, also known as lossless data hiding schemes were suggested to achieve the goal. For some applications, it requires to completely recover the host media if the marked images remain intact, and to extract the hidden message when the marked images were intentionally (or unintentionally) manipulated by the third parties. But, most of reversible data hiding schemes (Tian, 2003; Alattar, 2004; Hsio et al., 2009; Hu et al., 2009; Tai et al., 2009; Wu et al., 2009; Lee et al., 2010; Xiao & Shih, 2010; Yang & Tsai, 2010; Yang et al., 2010, 2011) were fragile in the sense that the hidden message can be unsuccessfully extract even if a slight alteration to the marked images, not to mention the recovery of the host media. Several authors (Zou et al., 2006; Ni et al., 2008; Zeng et al., 2010) therefore proposed robust reversible data hiding algorithms to overcome the issue.

Zou et al. (Zou et al., 2006) presented a semi-fragile lossless watermarking scheme based on integer wavelet transform (IWT). To obtain a good perceptual quality, they only embed data bits into the low-high (LH) and high-low (HL) of the IWT coefficients. During bit embedding, the IWT blocks remain intact if an input bit is 0, otherwise, the proposed embedding process were applied to the blocks. Simulations showed that the hidden message was robust against lossy compression to a certain degree. Ni et al. (Ni et al., 2008) presented a robust lossless data hiding technique based on the patchwork theory, the distribution features of pixel groups, error codes, and the permutation scheme. The marked images generated by the technique contained no salt-and-pepper noise with a limited payload size. In addition, the marked images were robust against to JPEG/JPEG2000 compression. Zeng et al. (Zeng et al., 2010) adjusted the mathematical difference values of a block and designed a robust lossless data hiding scheme. A cover image was first divided into a number of blocks and the arithmetic difference of each block was calculated. Data bits were then embedded into the blocks by shifting the arithmetic difference values. Due to the separation of the bit-0-zone and the bit-1-zone, as well as the particularity of mathematical difference, a major merit of the method was tolerant of JPEG compression to some extent. Compared with Ni et al.'s work (Ni et al., 2008), the performance of Zeng et al.'s scheme (Zeng et al., 2010) was significantly improved.

Currently there are a few robust lossless data hiding techniques published in the literature. Since the payload provided by the above techniques (Zou et al., 2006; Ni et al., 2008; Zeng et al., 2010) was not good enough, we therefore propose the FBBE algorithm so that to introduce an effective robust lossless data hiding method. Moreover, to provide a high-capacity version of lossless data hiding scheme that based on IWT domain, we use a smart allocation of the coefficients in an IWT block to achieve the goal. The scheme not only provides a high payload but also generates a good perceived quality.

This chapter is organized as follows. In section 2, a robust lossless data hiding via the feature-based bit embedding (FBBE) algorithm is introduced followed by a high-performance lossless data hiding scheme. Section 3 provides both test results and performance comparisons. We conclude this chapter in section 4.

2. Proposed method

Based on the integer wavelet transform (IWT), we propose two lossless data hiding methods, namely, a robust version and a high-capacity one. First, a robust lossless data hiding via the feature-based bit embedding (FBBE) algorithm is specified. Instead of embedding data bits directly into the IWT coefficient blocks, we use the FBBE algorithm to encode a block so that it can carry data bits and can be successfully identified later at the receiver. Then, a high-performance lossless data hiding scheme is presented to provide a large hiding storage by adjusting the location of each IWT coefficient in the host block. More specifically, the FBBE algorithm can be used to generate a robust lossless data hiding method. Whereas, the proposed smart adjustment of the IWT coefficients can be used to generate a high-performance lossless data hiding scheme.

2.1 FBBE algorithm

To achieve a robust lossless data hiding method, we embed a secret message into transform domain via the FBBE algorithm. An input image was first decomposed to the IWT domain. The IWT coefficients can be acquired by using the following two formulas:

$$d_{1,k} = s_{0,2k+1} - s_{0,2k} \qquad (1)$$

and

$$s_{1,k} = s_{0,2k} + \left\lfloor \frac{d_{1,k}}{2} \right\rfloor, \qquad (2)$$

where $s_{j,k}$ and $d_{j,k}$ are the kth low-frequency and high-frequency wavelet coefficients at the jth level, respectively (Calderbank et al., 1998). The $\lfloor x \rfloor$ is a floor function. Then, data bits were embedded into the blocks which derived from the LH and HL sub-bands of the IWT coefficients, respectively. The FBBE algorithm consists of four parts, namely, Up-U (UU) sampling, Down-U (DU) sampling, Up-Down (UD) sampling, and Left-Right (LR) sampling. Each sampling is allowed to carry a single data bit. For each host block, the above four samplings is conducted according to the sequence of UU, DU, UD, and LR samplings. The details are specified in the following sections.

2.1.1 Bit embedding

Let $C_j = \{c_{jk}\}_{k=0}^{n^2-1}$ be the jth block of size $n \times n$ taken from the LH (or HL) sub-bands of IWT domain. Also let $C_j = \{\hat{C} \cup \tilde{C} \cup C' \cup C''\}$ with $\hat{C} = \{\hat{c}_i \mid i = 0,3,5,6\}$, $\tilde{C} = \{\tilde{c}_u \mid u = 9,10,12,15\}$, $C' = \{c'_v \mid v = 1,2,13,14\}$, and $C'' = \{c''_w \mid w = 4,7,8,11\}$ be the UU, DU, UD, and LR samplings coefficients, respectively, as shown in Fig. 1 if $n=4$. In addition, let

$$C_{jp} = \{\hat{c}_i \mid \beta \le \hat{c}_i < 2\beta\} \qquad (3)$$

and

$$C_{jm} = \{\hat{c}_i \mid -2\beta \le \hat{c}_i < -\beta\} \qquad (4)$$

be the two focal groups being used to 'carry' data bits. The β used here is a robustness parameter.

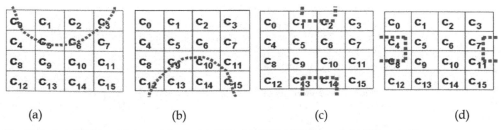

(a)　　　　　　　(b)　　　　　　　(c)　　　　　　　(d)

Fig. 1. A 4×4 IWT coefficients block. (a) UU, (b) DU, (c) UD, and (b) LR sampling coefficients.

The main steps of UU (or DU, UD, LR) samplings are specified as follows:

Step 1. Input a block C_j not processing yet.

Step 2. If an input bit $\phi=0$ and $|C_{jp}|>|C_{jm}|$ then do nothing, which means a bit 0 can be carried by the UU (or DU, UD, LR) sampling coefficients without alteration of their value, and go to Step 8.

Step 3. If $\phi=0$ and $|C_{jp}|=|C_{jm}|$ then add β to the coefficients c_{jk} in C_j with $0 \le c_{jk} < \beta$, respectively, mark a flag to the shifted coefficient, and go to Step 8.

Step 4. If $\phi=0$ and $|C_{jp}|<|C_{jm}|$ then add β to the coefficients in C_{jm}, respectively, mark a flag to the shifted coefficient, and go to Step 8.

Step 5. If $\phi=1$ and $|C_{jp}|<|C_{jm}|$ then do nothing, which means the UU (or DU, UD, LR) samplings coefficients carry a bit 1, and go to Step 8.

Step 6. If $\phi=1$ and $|C_{jp}|=|C_{jm}|$ then subtract β from the coefficients c_{jk} in C_j with $-\beta \le c_{jk} < 0$, respectively, mark a flag to the shifted coefficient, and go to Step 8.

Step 7. If $\phi=1$ and $|C_{jp}|>|C_{jm}|$ then subtract β from the coefficients in C_{jp}, respectively, mark a flag to the shifted coefficient.

Step 8. Repeat Step 1 until all IWT coefficients blocks have been processed.

Notice that the coefficients $\hat{c}_i$ which belong to either C_{jp} or C_{jm} have to be changed to $\tilde{c}_u$, c'_v, or c''_w, respectively, when the DU, UD, or LR samplings was employed. From the above procedures we can see that each block can carry at most four data bits. This resulted in a total payload of $\lfloor M/2n \rfloor \times \lfloor N/2n \rfloor \times 2 \times 4 \le \dfrac{2MN}{n^2}$ bits provided by the proposed method, where M and N is the size of a host image.

2.1.2 Bit extraction

Let $D_j = \{d_{jk}\}_{k=0}^{n^2-1}$ be the jth hidden block of size $n \times n$ taken from the LH (or HL) sub-bands of IWT domain derived from a marked image, and $D_j = \{\hat{D} \cup \tilde{D} \cup D' \cup D''\}$ with $\hat{D} = \{\hat{d}_i \mid i=0,3,5,6\}$ $\tilde{D} = \{\tilde{d}_u \mid u=9,10,12,15\}$ $D' = \{d'_v \mid v=1,2,13,14\}$, and $D'' = \{d''_w \mid w=4,7,8,11\}$. Also let

$$D_{jp} = \left\{\hat{d}_i \,(\text{or } \tilde{d}_u, d'_v, d''_w) \mid \beta \le \hat{d}_i \,(\text{or } \tilde{d}_u, d'_v, d''_w) < 2\beta\right\} \tag{5}$$

and

$$D_{jm} = \left\{\hat{d}_i \,(\text{or } \tilde{d}_u, d'_v, d''_w) \mid -2\beta \le \hat{d}_i \,(\text{or } \tilde{d}_u, d'_v, d''_w) < -\beta\right\} \tag{6}$$

be the two subsets of D_j. The procedure of bits extraction for the UU (or DU, UD, LR) sampling can be summarized in the following steps.

Step 1. Input a hidden block D_j not processing yet.

Step 2. If $|D_{jp}|>|D_{jm}|$ then a bit 0 can be identified. Subtract β from either the coefficients d_{jk} in D_j with $-\beta \le d_{jk} < 0$ or the coefficients $\hat{d}_i$ (or $\tilde{d}_u, d'_v, d''_w$) in D_{jp} when the corresponding flag was set at 1, and go to Step 6.

Step 3. If $|D_{jp}|<|D_{jm}|$ then a bit 1 can be extracted. Add β to either d_{jk} in D_j with $0 \le d_{jk} < \beta$ or the coefficients $\hat{d}_i$ (or $\tilde{d}_u, d'_v, d''_w$) in D_{jm} when the corresponding flag was set at 1, go to Step 6.

Step 4. If $|D_{jp}|=|D_{jm}|$ and the flag of the coefficients d_{jk} in D_j with $-\beta \le d_{jk} < 0$ was set at 1, a bit 0 can be identified, and go to Step 6.

Step 5. If $|D_{jp}|=|D_{jm}|$ and the flag of the coefficients d_{jk} in D_j with $0 \le d_{jk} < \beta$ was set at 1, a bit 1 can be identified.

Step 6. Repeat Step 1 until all hidden bits have been extracted.

The number of bits for the overhead information which used to signify whether or not a coefficient of the block undergone adjustment is $\left\lfloor \dfrac{M}{2}/n \right\rfloor \times \left\lfloor \dfrac{N}{2}/n \right\rfloor \times n^2 \times 2 \le \dfrac{MN}{2}$.

2.2 High-performance lossless data hiding scheme

To provide a high-capacity with a good perceived quality, the proposed scheme, which based on the adjustment of the locations of the coefficients in a host block, embeds a secret message into the three high sub-bands of IWT domain. The details are described in the following subsections.

2.2.1 Data embedment

Let $C_j = \{c_{jk}\}_{k=0}^{n^2-1}$ be the jth block of size $n \times n$ taken from the LH (or HL, HH) sub-band of IWT domain. Also let $C_{jp} = \{c_p | \beta \le c_p < 2\beta\}$ and $C_{jm} = \{c_m | -2\beta \le c_m < -\beta\}$ be two subsets of C_j. The main steps of bit embedding are specified as follows:

Step 1. Input a block C_j not processing yet.

Step 2. If $|C_{jp}| \ne \varnothing$ then subtract β from each coefficient of C_{jp} and mark a flag to the modified coefficient.

Step 3. If $|C_{jm}| \ne \varnothing$ then add β to each coefficient of C_{jm} and mark a flag to the modified coefficient.

Step 4. After adjustment, for a coefficient $c_i \in C_j$ with $0 \le c_i < \beta$ (or $-\beta \le c_i < 0$), multiply c_i by 2 to obtain $\hat{c}_i$, and add an input bit to $\hat{c}_i$.

Step 5. Repeat to Step 1 until all blocks have been processed.

The purpose of steps 3 and 4 are tried to further dig out hiding space from the selected coefficients. The schema of the adjustment of the coefficients values for the above two steps can be illustrated in Fig. 2.

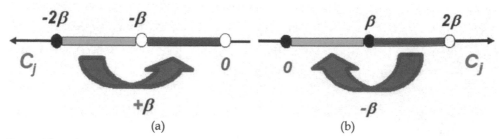

Fig. 2. The schema of the coefficients adjustment. (a) The positive part and (b) the negative part.

To increase payload size, multiple bits can be hidden in each IWT coefficient. In this case, the above steps 2-4 are rewritten as follows:

Step 2a. For a coefficient $c_t \in C_j$ with $-\beta < c_t < \beta$, multiply c_t by 2^k to obtain $\hat{c}_t$, and mark a flag to the modified coefficient.

Step 3a. For each $\hat{c}_t$, add data bits ϕ to $\hat{c}_t$ if $\hat{c}_t \geq 0$, otherwise, subtract ϕ from $\hat{c}_t$.

The parameter k is an integer. To maintain a good resulting perceived quality, the value of k is no more than 2. From the above procedure we can see that the number of bits used for recording the indices of the modified coefficients is $\lfloor M/2n \rfloor \times \lfloor N/2n \rfloor \times n^2 \times 3 < 3MN/4$.

2.2.2 Data extraction

To extract the hidden message, the overhead information can be losslessly compressed by using either the run-length coding algorithm or JBIG2. The resulting bit stream can then sent by an out-of-band transmission to the receiver. Without loss of generality, let D_j be the jth hidden block of size $n \times n$ taken from the LH (or HL, HH) sub-band of IWT domain which derived from a marked image, and $\hat{D}_j = \{\hat{d}_j | -2\beta \leq \hat{d}_j < 2\beta\}$ with $\hat{D}_j \subseteq D_j$. The procedure of bits extraction can be summarized in the following steps.

Step 1. Input a block D_j not processing yet.

Step 2. A data bit can be extracted by performing modulus-2 to $\hat{d}_j$.

Step 3. The IWT coefficients $\tilde{d}_j$ which hid data bit can be restored by performing either $\tilde{d}_j = \lfloor \hat{d}_j/2 \rfloor$ if $\hat{d}_j \geq 0$ or $\tilde{d}_j = \lceil (\hat{d}_j/2) - 0.5 \rceil$ if $\hat{d}_j < 0$.

Step 4. The original IWT coefficients can be recovered by adding (or subtracting) β to (or from) $\tilde{d}_j$ if $\tilde{d}_j \geq 0$ (or $\tilde{d}_j < 0$) while the flag of $\tilde{d}_j$ was marked.

Step 5. Repeat to Step 1 until all data bits have been extracted.

Note that $\lfloor x \rfloor$ and $\lceil x \rceil$ in step 3 denote the floor and ceiling functions, respectively. To perform multiple bits extraction for each coefficient, the above steps 2-4 are rewritten as follows:

Step 2b. A data bit can be extracted by performing modulus-2^k to d'_j with $-2^k \beta \le d'_j < 2^k \beta$.

Step 3b. The IWT coefficients $\tilde{d}_j$ which hid data bits can be restored by performing $\tilde{d}_j = \lfloor \hat{d}_j / 2^k \rfloor$ if the flag of $\tilde{d}_j$ was marked.

Step 4b. The original IWT coefficients can be recovered by adding (or subtracting) $(2^k - 1)\beta$ to (or from) $\tilde{d}_j$ if $\tilde{d}_j \ge 0$ (or $\tilde{d}_j < 0$).

To specify the idea of data embedding, two examples were presented in Figs. 3-4. The figures illustrate the cases of full-bit ($n{\times}n$ bits) and partial-bit hidden, respectively. The k used here is 1. The control parameter β is set to be 4. A host IWT-block was shown in Fig. 3(a). Figure 3(b) illustrates a shifted block, which obtained by according to the steps 2-3 of Sec. 2.2.1. Note that each of the shifted coefficients was marked by a rectangle. According to the step 4 of Sec. 2.2.1, we can see that all of the coefficients in the shifted block can be used to hide bits. Namely, a full-bit (or 16-bit) can be embedded in Fig. 3(b). Figure 3 (c) shows the hidden block. The mean square error (MSE) computed from Figs. 3(a) and 3(c) is 7.667. Another example of hiding partial-bit (or 12-bit) in an IWT-block was shown in Fig. 4(a). A shifted block was shown in Fig. 4(b). Notice as well there are 4 coefficients (in bold type) containing null bits. The resulting hidden block was depicted in Fig. 4(c). In this case, the MSE for the hidden block is 6.444. To recover the original block, a similar reverse process (with a bitmap) can be performed to Figs. 3(c) and 4(c), respectively.

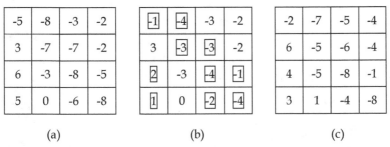

-5	-8	-3	-2
3	-7	-7	-2
6	-3	-8	-5
5	0	-6	-8

(a)

[-1]	[-4]	-3	-2
3	[-3]	[-3]	-2
[2]	-3	[-4]	[-1]
[1]	0	[-2]	[-4]

(b)

-2	-7	-5	-4
6	-5	-6	-4
4	-5	-8	-1
3	1	-4	-8

(c)

Fig. 3. Example of 16-bit embedding with a bit-stream of 0110 0100 0101 1100. (a) The original IWT-block, (b) shifted block, and (c) hidden block.

8	17	0	11
0	1	-15	2
-5	1	-6	-7
6	5	1	4

(a)

8	**17**	0	**11**
0	1	-15	2
[-1]	1	[-2]	[-3]
[2]	[1]	1	[0]

(b)

8	17	0	11
1	3	-15	4
-2	3	-3	-6
4	2	3	1

(c)

Fig. 4. Example of 12-bit embedding with a bit-stream of 0110 0110 0011. (a) The IWT-block, (b) shifted block, and (c) hidden block.

2.2.3 Overflow/underflow issues

An overflow/underflow can be occurred during bit embedding if a pixel value of the host image is a little either less than 255 or larger than 0. To overcome the overflow/underflow issues, a pixel-shifting approach can be performed in the spatial domain before data embedment. Namely, if a pixel value p in a host image satisfied either $p < \phi_1$ or $p > \phi_2$, p can be adjusted to a new value by adding to ϕ_1 or subtracting from ϕ_2. Both ϕ_1 and ϕ_2 are two predetermined threshold values.

3. Experimental results

Several greyscale images of size 512×512 were used as host images. A quarter of the host image *Lena* was used as the test data. To provide a variety of embedding rate, the value of the control parameter β is not fixed. Simulations generated by the proposed FBBE algorithm were first shown in the following subsection. Subsequently, a high-performance hiding scheme was examined.

3.1 Simulations of the FBBE algorithm

Figure 5 depicts the relationship between peak signal-to-noise ratio (PSNR) and robustness parameter β that generated by the proposed FBBE algorithm. The size of the block was 4×4. The figure indicated that the optimal PSNR value of 57.45 dB is achieved with β=1.

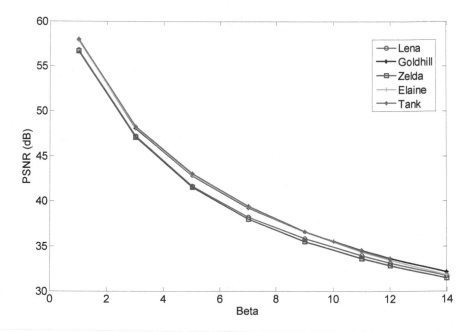

Fig. 5. The relationship between PSNR and β.

The PSNR value is approximately linear decreased as β increased. Actually, the larger the value of β, the more robust performance can be obtained by the proposed method. The PSNR is defined by

$$PSNR = 10 \times \log_{10} \frac{255^2}{MSE},$$ (7)

where $MSE = \frac{1}{MN} \sum_{i=1}^{N} \sum_{j=1}^{M} (\hat{x}(i,j) - x(i,j))^2$. Here $x(i,j)$ and $\hat{x}(i,j)$ denote the pixel values of the original image and the marked image. Figure 6 shows the marked images generated by the proposed method with β=12. Their average PSNR value was 33.35 dB with an embedding rate of 0.125 bits per pixel (bpp). It can be seen that the perceptual quality was acceptable.

Fig. 6. The marked images generated by the proposed FBBE algorithm. (a) Lena, (b) Goldhill, (c) Zelda, (d) Elaine, and (e) Tank.

For comparison, two graceful schemes, namely, Ni et al.'s algorithm (Ni et al., 2008) and Zeng et al.'s approach (Zeng et al., 2010) are compared with our method. Table 1 indicates the performance comparison of these methods on three test images. From Fig. 5 and Table 1 we can see that the proposed method with β =5 (or β of which value being less than 6) provides the largest payload among these methods while the PSNR for the proposed method is superior to that for the other two techniques. Moreover, Table 1 shows that the average hiding capacity provided by the proposed method is two times that achieved by Zeng et al.'s approach (Zeng et al., 2010), and five times larger than that achieved by Ni et al.'s algorithm (Ni et al., 2008).

Methods	Images			
	Lena	*Zelda*	*Goldhill*	Average
Ni et al.'s algorithm	6,336/ 40.19	4,480/ 40.47	6,336/ 40.18	5,717/ 40.28
Zeng et al.'s approach	16,384/ 38.07	16,384/ 38.09	16,384/ 38.10	16,384/ 38.09
Proposed Method	32,768/ 41.71	32,768/ 41.56	32,768/ 42.84	32,768/ 42.04

Table 1. Hiding performance (Payload/ PSNR) comparison between various methods.

To demonstrate the robustness performance of the proposed method, examples of extracted watermarks after various manipulations of the image are given in Table 2. A logo of size 63×63 with 8 bits/pixel 2 colours was used as the test watermark, as shown in Fig. 7. The bit correct ratio (BCR) is also included. The BCR is defined by

$$BCR = \left(\left. \frac{\sum\limits_{i=0}^{ab-1} \overline{w_i \oplus \widetilde{w}_i}}{a \times b} \right. \right) \times 100\%, \tag{8}$$

**CSIE
NPU**

Fig. 7. The test watermark.

where w_i and $\widetilde{w}_i$ represent the values of the original watermark and the extracted watermark respectively, as well as the size of a watermark is $a \times b$. Note that a majority-vote decision was employed during bits extraction. Although the BCR for those watermarks, which extracted from the images that gone through attacks such as JPEG2000, JPEG, equalized, interleaved, and inversion are not high, they are identifiable. Although the BCR for the watermark extracted from an image which manipulated by inversion attack is only 1.99%, it is recognizable. Furthermore, Fig. 8 shows the BCR performance of the survived watermarks under a variety of degree of Uniform/Gaussian noise additions attacks. From the figure we can see that the proposed method is more robust against Uniform than Gaussian noise additions attacks. Similarly, Fig. 9 indicates the proposed method has the better performance in resisting JPEG200 than JPEG attacks. Figure 10 shows that the proposed method is nearly free from brightness attacks. Finally, Fig. 11 indicates that the extracted watermarks are tolerant of colour quantization attack even if the number of level of pixel-value in a marked image is reduced to 8.

Attacks	Survived Watermarks	Attacks	Survived watermarks
Cropping (50%) BCR =87.88 %		Brightness (+90%) BCR = 87.45%	
JPEG2000 (CR*=8.33) BCR=71.89%		Brightness (-100%) BCR = 89.65%	
JPEG (CR=5.54) BCR=75.36%		Contrast (40%) BCR = 87.48%	
Uniform noise (5%) BCR = 78.94%		Contrast (-15%) BCR = 78.18%	
Gaussian noise (4%) BCR = 74.38%		Posterized (8-level) BCR = 85.26%	
Edge sharpening BCR = 98.92%		Equalized BCR = 80.78%	
Mean filtering (3×3) BCR = 98.34%		Interleaved (Odd) BCR = 54.14%	

Attacks	Survived Watermarks	Attacks	Survived watermarks
Median filtering (3×3) BCR = 98.76%	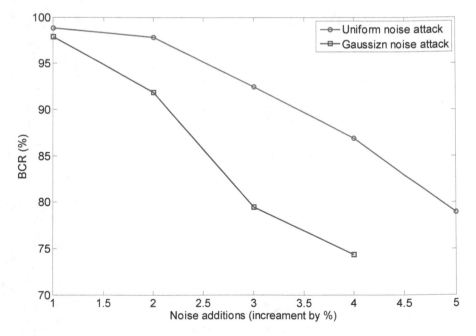	Interleaved (Even) BCR = 53.87%	
Quantization[+] BCR = 95.67%		Inversion BCR = 1.99%	

[*]CR stands for compression ratio, which is defined as the ratio of the size of a host image to that of a compressed image.
[+]The last four bits of the pixel in the marked image were truncated.

Table 2. Examples of watermarks extracted from image *Lena*. (β=12)

Fig. 8. The BCR for the proposed method under Uniform/Gaussian noise additions attacks, respectively.

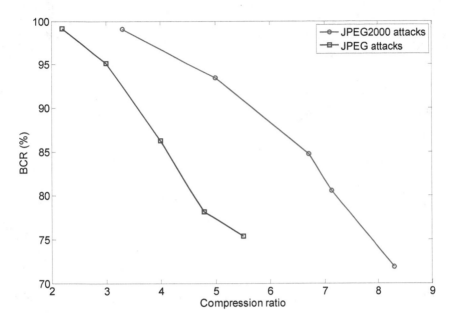

Fig. 9. The BCR for the proposed method under JPEG2000/JPEG attacks, respectively.

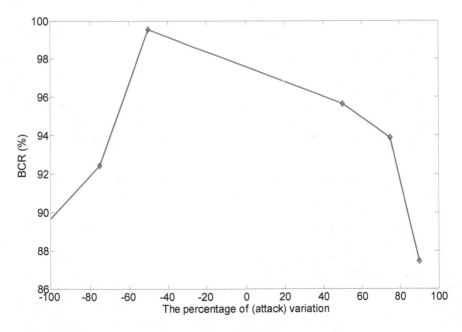

Fig. 10. The BCR for the proposed method under Brightness attacks.

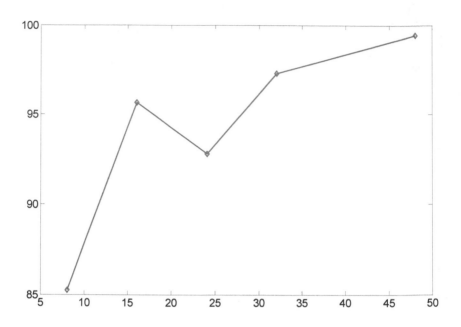

Fig. 11. The BCR for the proposed method under (color) quantization attacks.

3.2 Simulations of high-performance hiding scheme

The trade-off between PSNR and payload for the proposed scheme was depicted in Figure 12. The figure indicated that the average PSNR achieved by the proposed scheme was approximately 55 dB at a bit rate of 0.236 bpp. Whereas, the optimal PSNR value of 37.76 dB can be achieved in image *Zelda* with bit rate of 0.747 bpp. In addition, the relationship between payload (or embedding rate) and robustness parameter β was drawn in Fig. 13. From the figure we can see that the larger the value of β, the higher the bit rate was achieved.

For comparison, three outstanding approaches: Wu et al.'s scheme (Wu et al. 2009), Lee et al.'s algorithm (Lee et al., 2010), and Yang & Tsai's technique (Yang & Tsai, 2010) were compared with our method. Performance comparison between these methods was given in Table 3. It is obvious that the proposed method provides the largest payload among these methods while the PSNR for the proposed method is superior to that for the other three algorithms. Moreover, Table 3 implies that the hiding capacity provided by the proposed method is approximately two times that achieved by the Wu *et al.*'s scheme (Wu et al. 2009), and is two times that achieved by Lee et al.'s algorithm (Lee et al., 2010). Moreover, Table 4 revealed the superiority of our scheme when the PSNR value around 43 dB. The average embedding rate for the proposed scheme was two times larger than that for the Wu et al.'s technique (Wu et al. 2009).

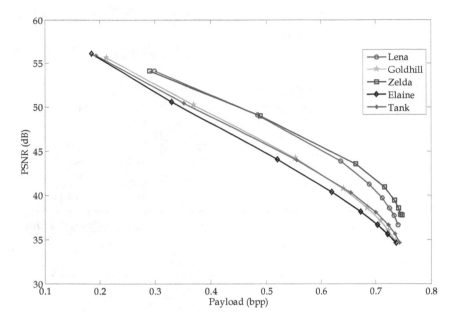

Fig. 12. The trade-off between payload and PSNR for the proposed scheme.

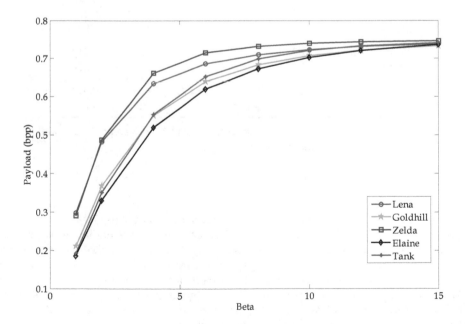

Fig. 13. The relationship between payload and β.

Methods	Images				
	Lena	*Zelda*	*Goldhill*	*Peppers*	Average
Wu et al.'s scheme	0.20/ 47.55	0.19/ 47.75	0.15/ 48.25	0.37/ 48.25	0.23/ 47.95
Lee et al.'s algorithm	0.23/ 48.25	0.18/ 48.25	-	0.17/ 48.25	0.20/ 48.25
Yang & Tsai's technique	0.38/ 48.81	-	0.26/ 48.81	0.33/ 48.81	0.33/ 48.81
Proposed method	0.48/ 49.14	0.49/ 49.02	0.40/ 50.27	0.43/ 49.37	0.45/ 49.45

Table 3. Embedding rate and PSNR performance comparison between various methods when PSNR value was approximately 48 dB.

Methods	Images				
	Lena	*Zelda*	*Goldhill*	*Peppers*	Average
Wu et al.'s scheme	0.24/ 43.60	0.40/ 43.60	0.28/ 43.60	0.23/ 43.60	0.29/ 43.60
Lee et al.'s algorithm	0.53/ 43.15	0.42/ 43.15	-	0.41/ 43.15	0.46/ 43.15
Yang & Tsai's technique	0.62/ 43.84	-	0.45/ 43.84	0.55/ 43.84	0.54/ 43.84
Proposed method	0.64/ 43.94	0.66/ 43.56	0.55/ 44.24	0.60/ 43.75	0.61/ 43.87

Table 4. Embedding rate and PSNR performance comparison between various methods when PSNR value was approximately 43 dB.

4. Conclusion

In this chapter, we first propose a robust lossless data hiding via the feature-based bit embedding (FBBE) algorithm based on integer wavelet transform (IWT). Data bits can be effectively carried by the IWT blocks via the FBBE algorithm and the hidden message can be successfully identified later at the receiver. Moreover, the FBBE algorithm can completely recover the host media if the marked image remains intact, and extract (most part of) the hidden message if manipulations were intentionally (or unintentionally) altered to the marked images. In addition, we employ a smart arrangement of the IWT coefficients so as to provide a high-capacity lossless data hiding scheme. Simulations validate that the marked images generated by the proposed FBBE algorithm are robust to a variety of attacks such as JPEG2000, JPEG, cropping, noise additions, (colour) quantization, bits truncation, brightness/contrast, mean/median filtering, and inversion. Furthermore, the payload and PSNR provided by the proposed two methods outperform those provided by existing schemes.

The proposed two methods can be extended to color images by embedding data bits in the RGB system separately. In addition, to further enlarge the hiding storage of the FBBE algorithm, an extra one (or two) data bits could be hidden in each IWT coefficients block during data embedment. However, a tradeoff between PSNR and payload size may be a problem with this algorithm. These issues will be discussed in detail in future work. Furthermore, to reduce memory space and transmission delay, the decreasing of the overhead bits will be our future study.

5. References

Alattar, A. M. (2004). Reversible watermark using the difference expansion of a generalized integer transform. *IEEE T. Image Processing*, Vol. 13, No. 8, pp. 1147-1156.

Al-Qaheri, H.; Mustafi, A. & Banerjee, S. (2010). Digital Watermarking using Ant Colony Optimization in Fractional Fourier Domain. *Journal of Information Hiding and Multimedia Signal Processing*, Vol. 1, No. 3, pp. 179-189.

Calderbank, A.R.; Daubechies, I.; Sweldens, W. & Yeo, B.L. (1998). Wavelet transforms that map integers to integers. *Applied & Computational Harmonics Analysis*, Vol. 5, No. 3, pp.332-369.

Cox, I.J.; Miller, M.L.; Bloom, J.A.; Fridrich, J. & Kalker T. (Ed(s.)) (2008). *Digital Watermarking and Steganography*, 2nd Ed., Morgan Kaufmann., MA.

Fan, L.; Gao, T. & Yang Q. (2011). A novel watermarking scheme for copyright protection based on adaptive joint image feature and visual secret sharing. *International Journal of Innovative Computing, Information and Control*, Vol. 7, No. 7(A), pp. 3679-3694.

Gu, Q. & Gao, T. (2009). A novel reversible watermarking algorithm based on wavelet lifting scheme. *ICIC Express Letters*, Vol. 3, No. 3 (A), pp. 397-402.

Hu, Y., Lee; H. K. & Li, J. (2009). DE-based reversible data hiding with improved overflow location map. *IEEE T. Circuits and Systems for Video Technology*. Vol. 19, No. 2, pp. 250-260.

Hsiao, J. Y.; Chan, K.F. & Chang, J.M. (2009). Block-based reversible data embedding. *Signal Processing*, Vol. 89, pp. 556-569.

Lai, C.C.; Huang, H.C. & Tsai, C.C. (2010). A digital watermarking scheme based on singular value decomposition and micro-genetic algorithm. *International Journal of Innovative Computing Information and Control*, Vol. 5, No. 7, pp. 1867-1873.

Lee, C.F.; Chen, H.L. & Tso, H.K. (2010). Embedding capacity raising in reversible data hiding based on prediction of different expansion. *The Journal of Systems and Software*, Vol. 83, pp. 1864-1872.

Lin, C.C. & Shiu, P.F. (2010). High capacity data hiding scheme for DCT-based images. *Journal of Information Hiding and Multimedia Signal Processing*, Vol. 1, No. 3, pp. 220-240.

Liu, J. C. & Shih, M. H. (2008). Generalization of pixel-value differencing staganography for data hiding in images. *Fundamenta Informaticate*, Vol. 83, pp. 319-335.

Martinez-Noriega, R.; Nakano, M.; Kurkoski, B. & Yamaguchi, K. (2011). High Payload Audio Watermarking: toward Channel Characterization of MP3 Compression. *Journal of Information Hiding and Multimedia Signal Processing*, Vol. 2, No. 2, pp. 91-107.

Ni, Z.; Shi, Y.Q.; Ansari, N.; Su, W.; Sun, Q. & Lin, X. (2008). Robust lossless image data hiding designed for semi-fragile image authentication," *IEEE T. Circuits and Systems for Video Technology*, Vol. 18, No. 4, pp. 497-509, 2008.

Qu, Z.G.; Chen, X.B.; Zhou, X.J.; Niu, X.X. & Yang, Y.X. (2010). Novel quantum steganography with large payload. *Optics Communications*, Vol. 283, No. 23, pp. 4782-4786.

Shih, F.Y. (2008). *Digital watermarking and steganography: fundamentals and techniques*. CRC Press, FL.

Tai, W.L.; Yeh, C.M. & Chang, C.C. (2009). Reversible data hiding based on histogram modification of pixel differences. *IEEE T. Circuits and Systems for Video Technology*, Vol. 19, No. 6, pp. 906-910.

Tian, J. (2003). Reversible data embedding using a difference expansion. *IEEE T. Circuits and Systems for Video Technology*, Vol. 13, No. 8, pp. 890-896.

Wang, S., Yang, B. & Niu, X. (2010). A secure steganography method based on genetic algorithm. *Journal of Information Hiding and Multimedia Signal Processing*, Vol. 1, No. 1.

Wu, H.C.; Lee, C.C.; Tsai, C.S.; Chu, Y.P. & Chen, H.R. (2009). A high capacity reversible data hiding scheme with edge prediction and difference expansion. *The Journal of Systems and Software*, Vol. 82, pp. 1966-1973

Xiao, D. & Shih, F.Y. (2010). A reversible image authentication scheme based on chaotic fragile watermark. *International Journal of Innovative Computing, Information and Control*, Vol. 6 No. 10, pp. 4731-4742.

Yamamoto, K. & Iwakiri M. (2010). Real-time audio watermarking based on characteristics of PCM in digital instrument. *Journal of Information Hiding and Multimedia Signal Processing*, Vol. 1, No. 2, pp. 59-71.

Yang, C.H. & Tsai, M.H. (2010). Improving histogram-based reversible data hiding by interleaving predictors. *IET Image Processing*, Vol. 4, No. 4, pp. 223-234.

Yang, C.Y.; Hu, W.C. & Lin, C.H. (2010). Reversible data hiding by coefficient-bias algorithm. *Journal of Information Hiding and Multimedia Signal Processing*, Vol. 1, No. 2, pp. 91-100.

Yang, C.Y.; Hu, W.C.; Hwang, W.Y. & Cheng, Y.F. (2010). A simple digital watermarking by the adaptive bit-labeling scheme. *Int. Journal of Innovative Computing, Information and Control*, Vol. 6, No. 3, pp. 1401-1410.

Yang, C.Y.; Lin, C.H. & Hu, W.C. (2011). Block-based reversible data hiding," *ICIC Express Letters*, Vol. 5, No. 7, pp. 2251-2256.

Zeng, X.T.; Ping, L.D. & Pan, X.Z. (2010). A lossless robust data hiding scheme. *Pattern Recognition*, Vol. 43, pp. 1656-1667.

Zhou, S.; Zhang, Q. & Wei, X. (2010). An image encryption algorithm based on dual DNA sequences for image hiding. *ICIC Express Letters*, Vol. 4, No. 4, pp. 1393-1398.

Zou, D.; Shi, Y.Q.; Ni, Z. & Su, W.A. (2006). A semi-fragile lossless digital watermarking scheme based on integer wavelet transform. *IEEE T. Circuits and Systems for Video Technology*, Vol. 16, No. 10, pp. 1294-1300.

10

Time-Varying Discrete-Time Wavelet Transforms

Guangyu Wang, Qianbin Chen and Zufan Zhang

*Chongqing Key Lab of Mobile Communications, Chongqing University of Posts and
Telecommunications (CQUPT)
China*

1. Introduction

Discrete-time wavelet transform (DWT) is found to be better than other transforms in the time-varying system analysis, e.g. for time-varying parametric modelling [16], time-varying systems identification [17], time-varying parameter estimation [18] and time domain signal analysis [19]. In the literature the common method to analyze the time-varying system using discrete-time wavelet transform is to model the time-varying system with a time-invariant system firstly, because a general analysis of time-varying discrete-time wavelet transform (TV-DWT) is still missing. To analyze the time-varying system directly using the time-varying discrete-time wavelet transform, we need the theory for the time-varying discrete-time wavelet transform.

The theory of time-invariant discrete-time wavelet transform (DWT) are quite complete [1,2,3]. For time-varying discrete-time wavelet transform, in literature there are some papers related with this topic by studying the changes of two different filter banks [10,11,12]. In [10] the authors analyzed the time-varying wavelet transform through changing the two-band filter banks used in the tree-structured implementation of DWTs with an simple example. In [11] the time-varying wavelet packets built with time-varying cosine-modulated filter banks were investigated. Similar with [10], in [12] the authors studied time-varying wavelet packets more theoretically with changing the two orthogonal two-band filter banks used in tree-structure of DWTs. Generally, in the existed theory of time-varying discrete-time wavelet transform it lacks a basic definition and description of the time-varying discrete-time wavelet transform. A basic analysis of time-varying discrete-time wavelet transform is also missing. The author has studied TV-DWT since some years and has published a series of papers about this topic. In this Chapter we summarize the author's main research results.

In our method the time-varying discrete-time wavelet transform is studied using a time-varying octave-band filter bank with tree structure. With this implementation the analysis of the time-varying discrete-time wavelet transform is equal to the analysis of the time-varying discrete-time octave-band filter bank. Then, the time-varying filter bank theory can be used in TV-DWT analysis. In this chapter we provide some theorems for the time-varying discrete-time wavelet transform with proofs.

2. Formulation of time-varying discrete-time wavelet transforms

From the point of view of digital signal processing, the time-varying discrete-time wavelet transform can be implemented by a time-varying octave-band filter bank with tree structure. Fig. 1 shows the most general time-varying discrete-time wavelet transform implemented with a time-varying octave-band filter bank, where the lowpass and highpass filter $H_l(z, m)$, $H_u(z, m)$, the stage number of the split-merge $J(m)$, all are varying with time index m. In other words, both the frequency characteristic and the time-frequency tiling of the discrete-time wavelet transform are varying with time. Fig. 2 shows the time-varying nonuniform filter bank implementation. With this implementation the analysis of the time-varying discrete-time wavelet transform is equal to the analysis of the time-varying discrete-time octave-band filter bank.

Note that we define the time-varying discrete-time wavelet transform varying with index m which is equivalent to the output index at the last stage of octave-band filter banks. The time indices of the other output are related to m by

$$m_j = 2^{J(m)-i-j} \cdot m, \ \ 0 \le j \le J(m) - 2. \tag{1}$$

In the literature there are some papers related with this topic by studying changes between two time-invariant filter banks [10,11,12]. In particular, in [10] the authors have discussed the transition behavior during the change between two time-invariant discrete-time wavelet transforms. Different from the existed publications, in this chapter we analyze the general time-varying discrete-time wavelet transform in detail based on the octave-band filter bank and the nonuniform filter bank implementation.

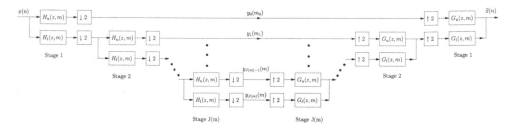

Fig. 1. Time-varying discrete-time wavelet transform implemented with time-varying octave-band filter banks.

3. Implementation with time-varying octave-band filter Banks

To make the analysis simple, in the following analysis we suppose that the stage number J does not change with time and is a constant. Then we get a J-stage time-varying octave-band filter bank. Just as depicted in Fig. 1, a J-stage octave-band time-varying filter bank consists of J stages of two-channel time-varying filter bank. In the analysis side, the input signal $x(n)$ is first split by the two-channel time-varying filter bank at the first stage, then the lowpass output is spilt again by the same two-band time-varying filter bank at the second stage.

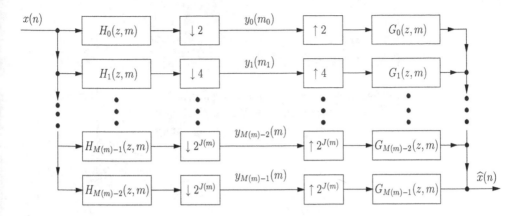

Fig. 2. Time-varying nonuniform filter bank implementation, where $M(m) = 2^{J(m)}$.

The process is ongoing until J-stage. In the synthesis side, the signal is merged to generate the reconstructed signal $\hat{x}(n)$. From the theorem of time-invariant discrete-time wavelet transform [2], we know that if the individual two-channel filter bank, or each split-merge pair is perfectly reconstructed, the octave-band filter bank is as well. Such statement is also valid for the time-varying octave-band filter bank. Therefore, we have following theorem.

Theorem 1: A time-varying discrete-time wavelet transform implemented with a time-varying octave-band filter bank is a biorthogonal time-varying transform if each two-channel time-varying filter bank is perfectly reconstructed.

We cannot use the method used in the time-invariant case to prove the above theorem because the system is time-varying. To prove theorem 1, we define analysis and synthesis matrices of the j-stage two-channel time-varying filter bank shown in Fig. 3 as

$$
\mathbf{T}_{ma}^{(j)} = \begin{bmatrix}
\vdots & \vdots & \vdots & \vdots & & \vdots & & \vdots \\
\cdots \mathbf{h}_0(m_j) & \mathbf{h}_1(m_j) & \cdots & \mathbf{h}_{N(j)-1}(m_j) & \mathbf{h}_{N(j)}(m_j) & \mathbf{h}_{N(j)+1}(m_j) & \cdots \\
\cdots \quad 0 & \mathbf{h}_0(m_j) & \cdots & \mathbf{h}_{N(j)-2}(m_j) & \mathbf{h}_{N(j)-1}(m_j) & \mathbf{h}_{N(j)}(m_j) & \cdots \\
\cdots \quad \vdots & \vdots & \ddots & \vdots & \vdots & \vdots & \cdots \\
\cdots \quad 0 & 0 & \cdots & \mathbf{h}_0(m_j) & \mathbf{h}_1(m_j) & \mathbf{h}_2(m_j) & \cdots \\
\cdots \quad 0 & 0 & \cdots & 0 & \mathbf{h}_0(m_j+1) & \mathbf{h}_1(m_j+1) & \cdots \\
& \vdots & \vdots & \vdots & \vdots & \vdots & \vdots
\end{bmatrix} \tag{2}
$$

$$\mathbf{T}_{ms}^{(j)} = \begin{bmatrix} \vdots & \vdots & \vdots & \vdots & \vdots \\ \cdots & g_0(m_j) & 0 & \cdots & 0 & 0 & \cdots \\ \cdots & g_1(m_j) & g_0(m_j) & \cdots & 0 & 0 & \cdots \\ \cdots & \vdots & \vdots & \ddots & \vdots & \vdots & \cdots \\ \cdots & g_{N(j)-1}(m_j) & g_{N(j)-2}(m_j) & \cdots & g_0(m_j) & 0 & \cdots \\ \cdots & g_{N(j)}(m_j) & g_{N(j)-1}(m_j) & \cdots & g_1(m_j) & g_0(m_j+1) & \cdots \\ \cdots & g_{N(j)+1}(m_j) & g_{N(j)}(m_j) & \cdots & g_2(m_j) & g_1(m_j+1) & \cdots \\ & \vdots & \vdots & & \vdots & \end{bmatrix}, \tag{3}$$

where

$$N(j) = 2^{J-j}, \quad j = 1, 2, \cdots, \tag{4}$$

and

$$\mathbf{h}_i(m_j) = \begin{bmatrix} h_u(L-2i-1, m_j) & h_u(L-2i-2, m_j) \\ h_l(L-2i-1, m_j) & h_l(L-2i-2, m_j) \end{bmatrix}, \tag{5}$$

$$\mathbf{g}_i(m_j) = \begin{bmatrix} g_u(2i, m_j) & g_l(2i, m_j) \\ g_u(2i+1, m_j) & g_l(2i+1, m_j) \end{bmatrix}, \tag{6}$$

where L is the filter length.

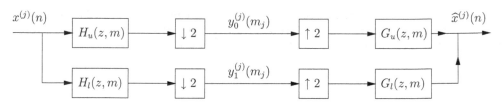

Fig. 3. The j-th stage two-channel time-varying filter bank.

Furthermore, we define two special matrices Λ_0 and Λ_1

$$\Lambda_0 = \begin{bmatrix} \vdots & \vdots & \vdots & \vdots & \vdots \\ \cdots & 1 & 0 & 0 & 0 & 0 & \cdots \\ \cdots & 0 & 0 & 1 & 0 & 0 & \cdots \\ \vdots & \vdots & \vdots & \vdots & \vdots \end{bmatrix} \tag{7}$$

$$\Lambda_1 = \begin{bmatrix} \vdots & \vdots & \vdots & \vdots & \vdots \\ \cdots & 0 & 1 & 0 & 0 & 0 & \cdots \\ \cdots & 0 & 0 & 0 & 1 & 0 & \cdots \\ \vdots & \vdots & \vdots & \vdots & \vdots \end{bmatrix}, \tag{8}$$

to extract the lowpass and highpass output like

$$y_0^{(j)} = \Lambda_0 y^{(j)}, \tag{9}$$

$$y_1^{(j)} = \Lambda_1 y^{(j)}, \tag{10}$$

where

$$\mathbf{y}^{(j)} = \left[\cdots y_0^{(j)}(-1) \; y_1^{(j)}(-1) \; y_0^{(j)}(0) \; y_1^{(j)}(0) \; y_0^{(j)}(1) \; y_1^{(j)}(1) \cdots \right]^T, \tag{11}$$

$$\mathbf{y}_0^{(j)} = \left[\cdots y_0^{(j)}(-1) \; y_0^{(j)}(0) \; y_0^{(j)}(1) \cdots \right]^T, \tag{12}$$

$$\mathbf{y}_1^{(j)} = \left[\cdots y_1^{(j)}(-1) \; y_1^{(j)}(0) \; y_1^{(j)}(1) \cdots \right]^T. \tag{13}$$

Based on the above matrix definitions we can describe the filter bank at the j-th stage showed in Fig. 3 as

$$\widehat{\mathbf{x}}^{(j)} = \mathbf{T}_{ms}^{(j)} \, \mathbf{T}_{ma}^{(j)} \, \mathbf{x}^{(j)}. \tag{14}$$

After adding the $(j+1)$-th stage with a biorthogonal time-varying two-channel filter bank shown in Fig. 4, we have

$$\widehat{\mathbf{x}}^{(j)} = \mathbf{T}_{ms}^{(j)} \, \mathbf{y}^{(j)}$$

$$= \mathbf{T}_{ms}^{(j)} \left\{ \Lambda_0^T \mathbf{y}_0^{(j)} + \Lambda_1^T \mathbf{y}_1^{(j)} \right\}$$

$$= \mathbf{T}_{ms}^{(j)} \left\{ \Lambda_0^T \Lambda_0 \mathbf{T}_{ma}^{(j)} \mathbf{x}^{(j)} + \Lambda_1^T \, \mathbf{T}_{ms}^{(j+1)} \, \mathbf{T}_{ma}^{(j+1)} \Lambda_1 \mathbf{T}_{ma}^{(j)} \mathbf{x}^{(j)} \right\}. \tag{15}$$

Because we suppose that the added two-channel filter bank is biorthogonal, we have

$$\mathbf{T}_{ms}^{(j+1)} \, \mathbf{T}_{ma}^{(j+1)} = \mathbf{I}, \tag{16}$$

$$\Lambda_0^T \Lambda_0 + \Lambda_1^T \Lambda_1 = \mathbf{I}. \tag{17}$$

Then, we can rewrite (15) as

$$\widehat{\mathbf{x}}^{(j)} = \mathbf{T}_{ms}^{(j)} \left\{ \Lambda_0^T \Lambda_0 + \Lambda_1^T \Lambda_1 \right\} \mathbf{T}_{ma}^{(j)} \, \mathbf{x}^{(j)}$$

$$= \mathbf{T}_{ms}^{(j)} \, \mathbf{T}_{ma}^{(j)} \, \mathbf{x}^{(j)}$$

$$= \mathbf{x}^{(j)} \tag{18}$$

which means that the time-varying octave-band filter bank is still perfectly reconstructed after adding next stage of time-varying biorthogonal two-channel filter bank. In other words, theorem 1 is correct.

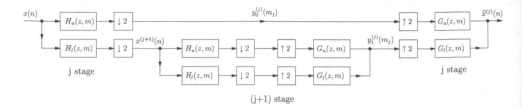

Fig. 4. Adding the $(j+1)$-th stage.

4. Implementation with time-varying nonuniform filter banks

Fig. 2 shows another implementation of a time-varying wavelet transform with $(J(m) + 1)$-channel time-varying nonuniform filter bank. To make the analysis easy we suppose that $J(m)$ does not change with time and is equal to constant J. For analysis of the $(J+1)$-channel time-varying nonuniform filter bank we first reconstruct the nonuniform filter bank to a time-varying uniform filter bank through adding following filters between $H_i(z,m)$ and $H_{i+1}(z,m)$ $(0 \leq j\,J - 1)$

$$H_{i,k}(z,m) = z^{-k \cdot 2^{i+1}} H_i(z,m), \ 1 \leq k \leq 2^{J-i-1} - 1. \tag{19}$$

After adding additional filters in the nonuniform filter bank in Fig. 2 the filter bank becomes M-channel time-varying uniform filter filter bank. The number of channel M is calculated by

$$M = \sum_{i=0}^{J-2} (2^{J-i-1} - 1) + (J+1)$$

$$= 2^{J-1} \sum_{i=0}^{J-2} 2^{-i} + 2$$

$$= 2^J (1 - 2 \cdot 2^{-J}) + 2$$

$$= 2^J. \tag{20}$$

For the time-varying system in Fig. 5 we have following theorem.

Theorem 2:   A time-varying discrete-time wavelet transform implemented with a time-varying nonuniform filter bank is biorthogonal if each two-channel time-varying filter bank in its tree-structured implementation is perfectly reconstructed.

To prove theorem 2, we need to describe the filter $H_i(z, m)$ in Fig. 5 based on the tree structure in Fig. 1. In the time-invariant discrete-time wavelet transform the description of such filters can be simply got using the convolution role in the transform-domain. However, in the time-varying case, we cannot describe $H_i(z, m)$ as product of functions in the previous stages, like $H_0(z, m)H_1(z^2, m)$, because the system is time-varying and the convolution role does not exist. Referencing to definitions of $\mathbf{T}_{ma}^{(j)}$ and $\mathbf{T}_{ms}^{(j)}$ in (2) and (3), we find that the analysis output

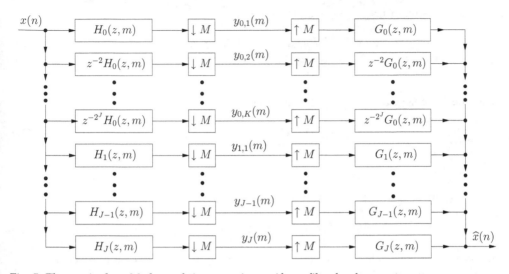

Fig. 5. The equivalent M-channel time-varying uniform filter bank.

$\mathbf{y}_0(m_0)$ can be expressed as

$$\mathbf{y}_0 = \boldsymbol{\Lambda}_0 \, \mathbf{T}_{ma}^{(1)} \, \mathbf{x}, \tag{21}$$

where

$$\mathbf{y}_0 = [\cdots \mathbf{y}_0(-1) \, \mathbf{y}_0(0) \, \mathbf{y}_0(1) \cdots]^T, \tag{22}$$

$$\mathbf{y}_0(m) = [\, y_{0,1}(m) \, y_{0,2}(m) \, \cdots y_{0,K}(m) \,]^T, \tag{23}$$

and $K = 2^{2^{J-1}} - 1$. In general, we have

$$\mathbf{y}_{j-1} = \underbrace{\boldsymbol{\Lambda}_0 \, \mathbf{T}_{ma}^{(j)} \, \boldsymbol{\Lambda}_1 \, \mathbf{T}_{ma}^{(j-1)} \, \cdots \, \boldsymbol{\Lambda}_1 \, \mathbf{T}_{ma}^{(1)}}_{\mathbf{H}_{j-1}} \, \mathbf{x}, \tag{24}$$

where $1 \le j \le J - 1$, and

$$\mathbf{y}_{J-1} = \boldsymbol{\Lambda}_0 \, \mathbf{T}_{ma}^{(J)} \, \boldsymbol{\Lambda}_1 \, \mathbf{T}_{ma}^{(J-1)} \, \cdots \, \boldsymbol{\Lambda}_1 \, \mathbf{T}_{ma}^{(1)} \, \mathbf{x} = \mathbf{H}_{J-1} \, \mathbf{x}, \tag{25}$$

$$\mathbf{y}_J = \boldsymbol{\Lambda}_1 \, \mathbf{T}_{ma}^{(J)} \, \boldsymbol{\Lambda}_1 \, \mathbf{T}_{ma}^{(J-1)} \, \cdots \, \boldsymbol{\Lambda}_1 \, \mathbf{T}_{ma}^{(1)} \, \mathbf{x} = \mathbf{H}_J \, \mathbf{x}, \tag{26}$$

where

$$\mathbf{y}_{j-1} = \left[\cdots \mathbf{y}_{j-1}(-1) \, \mathbf{y}_{j-1}(0) \, \mathbf{y}_{j-1}(1) \cdots \right]^T, \tag{27}$$

$$\mathbf{y}_{j-1}(m) = \left[y_{j-1,1}(m) \, y_{j-1,2}(m) \, \cdots y_{j-1,K}(m) \right]^T, \tag{28}$$

for $K = 2^{2^J - j} - 1$ and $1 \leq j \leq J - 2$, and

$$\mathbf{y}_{J-1} = \begin{bmatrix} \cdots & y_{J-1}(-1) & y_{J-1}(0) & y_{J-1}(1) & \cdots \end{bmatrix}^T, \tag{29}$$

$$\mathbf{y}_J = \begin{bmatrix} \cdots & y_J(-1) & y_J(0) & y_J(1) & \cdots \end{bmatrix}^T. \tag{30}$$

At synthesis side, we have similar definitions as

$$\widehat{\mathbf{x}}_{J-1} = \underbrace{\mathbf{T}_{ms}^{(1)} \boldsymbol{\Lambda}_1^T \mathbf{T}_{ms}^{(2)} \cdots \boldsymbol{\Lambda}_1^T \mathbf{T}_{ms}^{(j)} \boldsymbol{\Lambda}_0^T}_{\mathbf{G}_{j-1}} \mathbf{y}_{j-1}, \tag{31}$$

$$\widehat{\mathbf{x}}_{J-1} = \mathbf{T}_{ms}^{(1)} \boldsymbol{\Lambda}_1^T \mathbf{T}_{ms}^{(2)} \cdots \boldsymbol{\Lambda}_1^T \mathbf{T}_{ms}^{(J)} \boldsymbol{\Lambda}_0^T \mathbf{y}_{J-1} = \mathbf{G}_{J-1} \mathbf{y}_{J-1}, \tag{32}$$

$$\widehat{\mathbf{x}}_J = \mathbf{T}_{ms}^{(1)} \boldsymbol{\Lambda}_1^T \mathbf{T}_{ms}^{(2)} \cdots \boldsymbol{\Lambda}_1^T \mathbf{T}_{ms}^{(J)} \boldsymbol{\Lambda}_1^T \mathbf{y}_J = \mathbf{G}_J \mathbf{y}_J, \tag{33}$$

Now, based on the definition in (23), we can build the analysis output vector for th time-varying filter bank in Fig. 5 as

$$\mathbf{y} = \begin{bmatrix} \cdots & \mathbf{y}_0(-1) & \cdots \mathbf{y}_J(-1) & \mathbf{y}_0(0) & \cdots \mathbf{y}_J(0) & \mathbf{y}_0(1) & \cdots \end{bmatrix}^T. \tag{34}$$

Suppose that $\mathbf{T}_{ma}$ and $\mathbf{T}_{ms}$ are the analysis and synthesis matrices for the time-varying filter bank in Fig. 5. Referencing (34), $\mathbf{T}_{ma}$ is constructed by interleaving the rows from $\mathbf{T}_{ma}^{(1)}$ to $\mathbf{T}_{ma}^{(J)}$ with same time index m, $\mathbf{T}_{ms}$ is built with similar way, but interleaving the columns. Then, the production $\mathbf{T}_{ma}^{(1)} \mathbf{T}_{ma}^{(J)}$ can be expressed by

$$\mathbf{T}_{ms} \mathbf{T}_{ma} = \sum_{j=0}^{J} \mathbf{G}_j \mathbf{H}_j. \tag{35}$$

Substituting $\mathbf{H}_i$ and $\mathbf{G}_i$ defined in (24)-(26) and (31)-(33) into (35), and using properties in (16) and (17), we get

$$\mathbf{T}_{ms} \mathbf{T}_{ma} = \mathbf{I}, \tag{36}$$

which means that the time-varying nonuniform filter bank in Fig. 2 is perfectly reconstructed.

Finally, we give another property related with filter coefficients of the time-varying filter bank in Fig. 2. Suppose that $h_i(n, m)$ and $g_i(n, m)$ represent the analysis and synthesis filter coefficients in Fig. 2, then we have following equation

$$< g_i(n - kM, m + r), \; h_j(n - lM, m + s) > = \delta(k - l)\, \delta(i - j)\, \delta(r - s), \tag{37}$$

where $M = 2^J$. The proof of equation (37) can be simply got by using the PR condition in (36).

5. Conclusion

In the theory of discrete-time signal expansion, the wavelet transform is very important. In this chapter, we defined the general discrete time-varying dyadic wavelet transform and analyzed its properties in detail. Some theorems describing properties of time-varying discrete-time wavelet transforms were presented. The conditions for a biorthogonal time-varying discrete-time wavelet transform were given. The theory and algorithms presented in this chapter can be used in design of time-varying discrete-time signal expansion in practice.

6. Acknowledgments

This work is supported by the National Natural Science Foundation of China (No. 61071195) and Sino-Finland Cooperation Project (No. 1018).

7. References

[1] H.S. Malvar, *Signal Processing with Lapped Transforms.* Boston, MA: Artech House, 1992.

[2] M. Vetterli and J. Kovacevic, *Wavelets and Subband Coding,* Englewood Cliffs, NJ: Prentice Hall, 1995.

[3] Alfred Mertins, *Signal Analysis, Wavelets, Filter Banks, Time-Frequency Transforms and Application* England: John Wiley & Sons, 1999.

[4] V. DeBrunner, W. Lou, and J. Thripuraneni, "Multiple Transform Algorithms for Time-Varying Signal Representation," *IEEE Trans. Circuits and Systems-II: Analog and Digital Signal Processing,* vol. 44, No. 8, pp. 663-667, Aug. 1997.

[5] M. Morhac and V. Matousek, "New Adaptive Cosine-Walsh Transform and its Application to Nuclear Data Compression,"*IEEE Trans. Signal Processing,* vol. 48, No. 9, pp. 2663-2696, September 2000.

[6] N. Prelcic, A. Pena, "An Adaptive Tiling of the Time-Frequency Plane with Application to Multiresolution-Based Perceptive Audio Coding," *Signal Processing,* vol. 81, pp. 301-319, Feb. 2001.

[7] P. Srinivasan and L. Jamieson, "High-Quality Audio Compression Using an Adaptive Wavelet Packet Decomposition and Psychoacoustic Modeling," *IEEE Trans. Signal Processing,* vol. 46, No. 4, pp. 1085-1093, April 1998.

[8] ISO/IEC JTCI/SC29, "Information technology-coding of moving pictures and associated audio for digital storage media at up to about 1.5 Mbit/s-IS 11172-3 (audio)," 1992.

[9] C. Guillemot, P. Rault, and P. Onno, "Time-invariant and time-varying multirate filter banks: application to image coding ," *Annales des Telecommunications-Annals of Telecommunications,* vol. 53, No.5-6, pp.192-218, May-Jun 1998.

[10] I. Sodagar, K. Nayebi, and T.P Barnwell,III, "Time-varying filter banks and wavelets," *IEEE Trans. Signal Processing,* vol. 42, No. 11, pp. 2983-2996, Nov. 1994.

[11] R.L. de Queiroz, K.R. Rao, "Time-varying lapped transform and wavelet packets," *IEEE Trans. Signal Processing,* vol. 41, No. 12, pp. 3293-3305, Dec. 1993.

[12] C. Herley, M. Vetterli, "Orthogonal Time-Varying Filter Banks and Wavelet Packets," *IEEE Trans. Signal Processing,* vol. 42, No.10, pp. 2650-2663, Oct. 1994.

[13] G. Wang and U. Heute, "Time-varying MMSE modulated lapped transform and its applications to transform coding for speech and audio signals," *Signal Processing*, vol. 82, No. 9, pp. 1283-1304, Sept. 2002.

[14] G. Wang, "The most general time-varying filter bank and time-varying lapped transforms," *IEEE Trans. on Signal Proccesing*, vol. 54, No. 10, pp. 3775-3789, Oct. 2006.

[15] G. Wang, "Analysis of M-channel time-varying filter banks," *Digital Signal Processing*, vol. 18, No. 2, pp. 127-147, May 2008.

[16] H. Wei, S.Billings and J. Liu, "Time-varying parametric modelling and time-dependent spectral characterisation with applications to EEG signals using multiwavelets", *International Journal of Modelling, Identification and Control*, vol. 9, No. 3, pp. 215-224,2010.

[17] X. Xu, Z.Y. Shi and Q. You, "Identification of linear time-varying systems using a wavelet-based state-space method", *Mechanical Systems and Signal Processing*,doi:10.1016/j.ymssp.2011.07.005.

[18] Z. Lua, O. Gueganb, "Estimation of Time-Varying Long Memory Parameter Using Wavelet Method", *Communications in Statistics - Simulation and Computation*, vol. 40, No. 4, pp. 596-613, March 2011.

[19] M. Gokhale1 and D. Khanduja, "Time Domain Signal Analysis Using Wavelet Packet Decomposition Approach", *Int. J. Communications, Network and System Sciences*, doi:10.4236/ijcns.2010.33041.

[20] Guangyu Wang, Zufan Zhang, and Qianbin Chen, "Analysis and Properties of Time-Varying Modified DFT Filter Banks," *EURASIP Journal on Advances in Signal Processing*, vol. 2010, Article ID 342865, 15 pages, 2010. doi:10.1155/2010/342865.

[21] G. Wang, Q. Chen and Z. Ren, "Modelling of time-varying discrete-time systems", *IET Signal Processing*, vol. 5, No. 1, pp104-112, Feb. 2011.

Application of Wavelet Analysis for the Understanding of Vortex-Induced Vibration

Tomoki Ikoma, Koichi Masuda and Hisaaki Maeda

Department of Oceanic Architecture and Engineering,
College of Science and Technology (CST), Nihon University
Japan

1. Introduction

1.1 Marine riser

Oceans are quite important fields for us because many resources lurk there which are oil and gas under seabed, mineral resources, water heat energy and so on. Development of submarine oil has been major in the North Sea and in the Gulf of Mexico. Today, submarine oil has been developed at ultra-deep water fields of offshore of Brazil and West Africa, which does deep over 1000m. Deepest field developed is more than 3000m in water depth of Brazilian seas. Riser system is necessary to develop and to production submarine oil. The riser is a tubing structure which is for drilling and production. Diameter of a drilling riser is greater than 50cm and that of a production riser is about 20cm. The riser is thin rope-like tube in oceans. Therefore, the tubing behaves elastically by marine currents and ocean waves and so on.

These motion behaviors are called as Vortex-Induced Vibration (VIV). VIV of the riser is a complex phenomenon, which is dominated by the natural frequency of the riser system and behavior of vortex shedding around the rider. VIV is very important for structural design of the riser system and the platform of the drilling and the production of submarine oil and so on.

There are many studies of VIV of the riser and the drilling or the production system including the riser system in the ocean engineering field with numerical approaches, theoretical approaches and model experimental approaches. Behaviors of time variation of VIV obtained from numerical calculations or model experiments using model risers in a water tank include a complicated mechanism so it is not easy to understand them, because the time variation is not steady but transient and chaos. Therefore, we need to understand VIV phenomenon in not only time characteristics but also frequency characteristics.

For understanding frequency characteristics, we often use a power spectrum with the FFT analysis or others. However, a power spectrum does not inform us time variation of VIV characteristics. Then, the wavelet analysis can be applied to the VIV analysis because we can simultaneously understand the characteristics in time domain and frequency domain.

1.2 Application of wavelet analysis for study of marine riser

The authors have investigated VIV characteristics of a circular cylinder with forced oscillation tests in still water (Ikoma & Masuda et al., 2006, 2007). As these results, VIV behaviors have been classified to the four power spectrum pattern. However an actual orbit of the model cylinder was different even if the spectrum pattern was same. Therefore detail of VIV characteristics and behaviors cannot be understood from only a power spectrum with the FFT analysis of a time history of vibrations. In addition, a vibration phenomenon of a marine riser etc. is a non-steady problem in practice so that fluid velocity in the ocean and oscillation of an upper structure such like a production platform are an unsteady phenomenon. Therefore vibration characteristics such like VIV varies to time table.

The Hilbert transform was applied to analysis of cylinder vibration with VIV (Khalak & Wiliamson, 1999). In there, it is described that phase deviation occurs in region entering into VIV lock-in. The Hilbert transform was useful in order to analysis of marine riser vibrations and examined frequency characteristics which vary to time table.

The wavelet transform is applied to analysis of vibration problems with VIV of a rigid circular cylinder which cross-flow vibration is allowed due to vortex shedding in this study. The wavelet analysis is possible to do the time-frequency analysis as same as the Hilbert transform analysis. Objectives of this study are: 1) to examine possibility of application of the wavelet transform to VIV analysis and 2) to discuss VIV characteristics from results of the wavelet analysis. In 2010, the wavelet analysis and the Hilbert transform were also applied to the estimation of riser behaviors (Shi et al., 2010).

This chapter introduces application of the wavelet analysis in the ocean engineering field using results of VIV characteristics. From the model experiment, relationship between the orbit pattern of vibration of the model cylinder and a contour pattern of the wavelet is considered. As a result, effectiveness of the wavelet analysis in order to understand VIV detail is given.

2. Model experiment

2.1 Method of experiment

Model experiments using a single circular cylinder or two arranged circular cylinders in tandem are carried out at a wave tank that has 27 m in length, 7 m in width and 1 m in water depth in the campus of Funabashi at CST of Nihon University. We cannot generate current so that forced oscillation tests in still water are carried.

An experimental method and concepts follow our own past model testing (Ikoma & Masuda et al., 2006, 2007). In this study, a single cylinder or double cylinders in tandem arrangement to inline direction are used.

Test models of a cylinder are made of acryl resin which is rigid. However the cylinder system is not fixed because elastic vibration is allowed in only cross-flow direction by attaching a flat spring on top of the cylinder. The flat spring does not allow inline movement of the cylinder. Inline movement is due to only forced oscillation by the oscillator. VIV occurs such like rolling motion around x axis which center is the flat spring.

Number of degrees of freedom is two in sway motion and roll motion. The sway in this experiment is horizontal displacement of center of gravity in y direction and rolling is rotation around x axis in a coordinate system of Fig. 1. Freedom of surge corresponding to x axis motion is allowed and decided due to forced oscillation by experimental operators.

2.2 Experimental setup system

The cylinder is suspended under a load cell through the flat spring. Most of the cylinder is submerged in still water. Side views of the experimental setup system, which corresponds to the z-x plane, are illustrated in Fig. 2. The direction of forced oscillations is right and left in Fig. 2.

A load cell for measuring the total load in inline direction, which including the inertia force of a cylinder and hydrodynamic forces on a cylinder, is attached under the forced oscillation device. A Doppler current meter is installed at the straightly back of the cylinder, the current meter which moves together with the cylinder due to the forced oscillation. The current meter can measure fluid velocity of three directions in x, y and z axis. We measure the vertical bending moment with a flat spring on which strain gages are set.

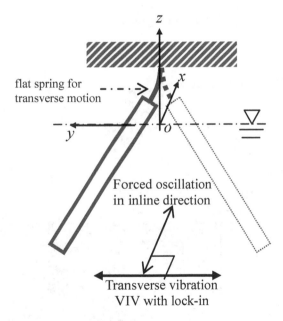

Fig. 1. Idealization of VIV in experiment

In the experiment, 1) inline displacement of forced oscillation with a potentiometer, 2) the total inline load with the load cell, 3) the bending moment with the flat spring and 4) fluid velocity at the back of the cylinder with the Doppler current meter are measured. The fluid velocity is measured at midship depth of the submerged cylinder. The VIV is evaluated by using the vertical bending moment and cross-flow displacement predicted from the bending moment.

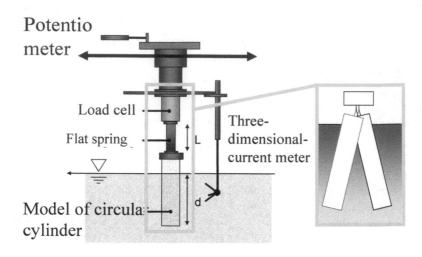

a)   in case of one cylinder

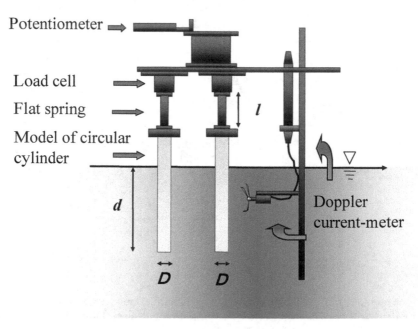

b)   in case of two cylinders

Fig. 2. Side views of experimental setup system

Photo 1. Side view of experimental setup system

2.3 Experimental conditions

Detail of the cylinder models is described in the paper (Ikoma et al., 2007). Length of the flat spring is expressed as "l" in Table 1. Natural periods T_n of cross-flow vibration of a suspended cylinder were obtained with the plucked decay test in still water.

Water depth is set to 1.0 m. The amplitude of forced oscillation is 7.2 cm, the Keulegan-Carpenter (K_C) number accordingly corresponds to 5.7 and 9.0 in the experiments.

In case of double cylinders, the cylinders are straightly suspended, and then the distance l_d between the center to center of both the cylinders is varied such as Table 2. The distance ratio s is defined as follows,

$$s = \frac{l_d}{D}.$$

(1)

The front cylinder and the back cylinder are defined as Fig. 3.

Photo 2. Experimental models filled with sand

case	diameter D	draft d	l	filled with water		filled with sand	
				measured natural period in water Tn	period of forced oscillation in case of $St=0.2$, Ts	measured natural period in water Tn	period of forced oscillation in case of $St=0.2$, Ts
1	8cm	30cm	10cm	0.86s	0.95s	0.93s	1.05s
2	8cm	80cm	13cm	3.28s	3.70s	2.72s	3.10s
3	5cm	30cm	10cm	0.54s	1.15s	0.57s	1.05s
4	5cm	80cm	13cm	2.15s	3.90s	2.08s	3.75s
5	5cm	80cm	10cm	1.86s	3.35s	1.85s	3.35s
6	5cm	60cm	10cm	1.21s	2.20s	1.28s	2.30s
7	8cm	60cm	10cm	1.91s	2.15s	1.79s	2.00s
8	8cm	60cm	4cm	1.18s	1.35s	1.20s	1.35s
9	8cm	80cm	4cm	1.91s	2.15s	1.83s	2.10s
10	5cm	80cm	4cm	1.20s	2.20s	1.23s	2.25s

a) for single cylinder

model	diameter D	draft d	l	measured natural period in water Tn	period of forced oscillation in case of $St=0.2$, Ts
1	5cm	60cm	10cm	1.28 s	2.00 s
2	5cm	80cm	10cm	1.85 s	3.35 s
3	8cm	60cm	10cm	1.79 s	2.00 s
4	8cm	80cm	10cm	2.52 s	2.85 s

b) for double cylinders

Table 1. Principal particulars of cylinder model setting

D	5 cm					8 cm	
l_d cm	10	13	15	18	20	16	20
S	2.0	2.5	3.0	3.5	4.0	2.0	2.5

Table 2. Variation of distance ratio of straight cylinders

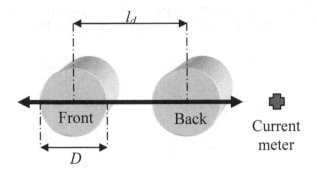

Fig. 3. Distance ratio between two cylinders

2.4 Definitions of nominal period and nominal frequency

The K-C number and the period of T_s are defined as same as them (Ikoma et al., 2007) as follows,

$$K_c = \frac{U_o T}{D},\qquad(2)$$

where U_O is the maximum velocity of forced oscillation, T stands for the period of the forced oscillation. The forced oscillation is simple harmonic motion in this study, hence eqn. (2) can be rewritten as follows,

$$K_c = 2\pi \frac{a}{D},\qquad(3)$$

in which a is amplitude of the forced oscillation in inline direction.

The range of periods of the forced oscillation is from about 0.4 seconds to 4.6 seconds, the Reynolds (R_e) numbers accordingly correspond to about 5.0e+3 to 6.0e+5 if the maximum velocity U_O of the forced oscillations are applied to the calculation.

The natural frequency of transverse vibration varies due to the length of a flat spring. The experimental conditions of each case are shown in Table 1. 'S_t' in Table 1 is the Strouhal number and is defined in this study as follows,

$$S_t = \frac{f_s D}{U_o},\qquad(4)$$

in which f_s is the frequency of vortex shedding. The Strouhal number has been well known as about 0.2 in range of R_e>1.0e+3. In this paper, the Strouhal number is approximately fixed to,

$$S_t \approx 0.2 \,. \tag{5}$$

'T_s' in Table 1 corresponds to the period of the forced oscillation which corresponds to about 5.0 in the nominal reduced velocity. The lock-in phenomenon of VIV is therefore expected in each experimental case when the model is in forced inline oscillations with the period of T_s. 'T_s' is calculated with following equations,

$$\frac{f_s D}{U_o} = 0.2 \,, \tag{6}$$

$$f_s = 0.2 \frac{U_o}{D} \,, \tag{7}$$

$$T_s = \frac{1}{f_s} \,. \tag{8}$$

Therefore, the frequency of vortex shedding f_s is not an actual frequency, but is a nominal frequency in this study.

3. Wavelet analysis

The wavelet analysis is the time-frequency analysis for time histories such like the Hilbert transform. The wavelet transform is defined as follows,

$$W_T(b,a) = \frac{1}{\sqrt{|a|}} \int_{-\infty}^{\infty} f(t) \cdot \psi\left(\frac{t-b}{a}\right) dt \,, \tag{9}$$

where $f(t)$ is a time history, a stands for a dilation parameter and b stands for a location parameter. "$\psi(t)$" is the mother wavelet function. The Gabor's mother wavelet is applied such as follows in this study,

$$\psi(t) = \frac{1}{\sqrt{2\pi\sigma^2}} \exp\left[-\frac{t^2}{2\sigma^2}\right] \cdot e^{i\omega_0 t} \,, \tag{10}$$

where σ is a damping parameter of the mother wavelet function and ω_0 is a principle angular frequency. Half of the natural angular frequency of each model is applied in this study.

When the damping parameter σ becomes smaller, the mother function be attenuated soon. And then, resolution of frequency is high although resolution of time gets worse. It is a merit to select the Gabor's wavelet because the dilation parameter a, which corresponds to a scaling parameter, is individual to the resolution parameter σ. The parameters are consequently individual each other so that the tuning of the parameters in order to draw the wavelet contour is not difficult.

In this study, b is set at 0.4 seconds, a is carried out from 0.0 to 3.0 with resolution of 0.2 and σ is 1.0.

A sampling frequency of the experimental recording has been 500 Hz, which corresponds to 2.0e-3 seconds in the sampling time. '0.4 seconds' of b in the time sifting for the wavelet analysis corresponds to 200 sampling data skipping. In addition, the shortest natural period of cross-flow vibration in the experimental models in Tables 1a) and 1b) is 0.86 seconds. If the bi-harmonic vibration in VIV in this case occurs, the vibration period is 0.43 seconds. The resolution would be thereby enough 0.4 seconds. Using $b=0.4$, variation pattern of the wavelet would be able to be reproduced. The step of a is now 0.2. The parameter a corresponds to a resolution of the frequency component. The cross-flow vibration appears relatively simply from the FFT analysis so that the resolution of 0.2 may be reasonable.

4. Orbit patterns and power spectrum patterns

In case of a single cylinder experiment (Ikoma & Masuda et al., 2006, 2007), the four patterns of the power spectrum of VIV have been found such like Fig. 4. In addition, there was an adequate correlation between the orbit pattern and the spectrum pattern in the paper (Ikoma et al., 2007). However both the power spectrum patterns of the orbit patterns of the type U and the type 8 correspond to the pattern 4 which is bi-harmonic type. Therefore detail of VIV behavior cannot be understood from a result with the FFT analysis of VIV.

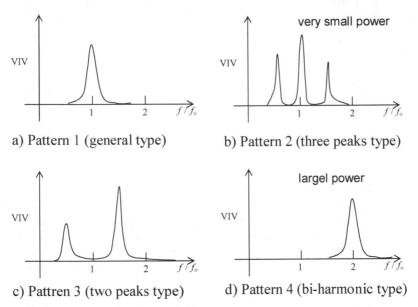

a) Pattern 1 (general type)   b) Pattern 2 (three peaks type)

c) Pattren 3 (two peaks type)   d) Pattern 4 (bi-harmonic type)

Fig. 4. Classifications of power spectrum patterns of VIV [1]

5. Results and discussion

5.1 Orbit patterns

From the experiment using the single cylinder, orbit patterns can be classified to six patterns such as Fig. 5. The net type is specified to $N1$ and $N2$. It can be considered that response of the type U and the type 8 corresponds to VIV lock-in.

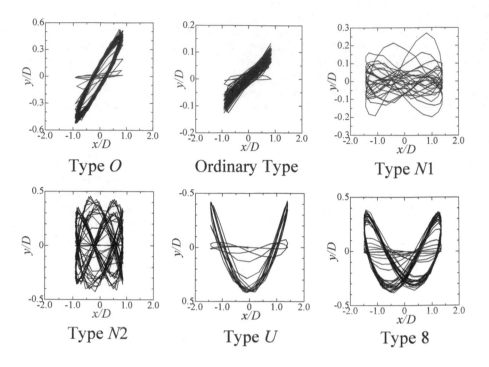

Fig. 5. Classifications of orbit patterns of VIV

5.2 Wavelet characteristics of single cylinder

In case of a single cylinder, wavelet patterns are made a general classification to five. These patterns are not decided due to experimental cases such as Table 1-a). Results of wavelet analysis in case of the case 8 show in Figs. 6, where "T" in subtitles stands for period of the forced oscillation. A vertical axis is the dilation parameter a and horizontal axis is time of VIV response. These results are the vertical bending moment which corresponds to cross-flow vibration.

In Figs. 6, a) corresponds to the orbit pattern of Type O. In Wavelet contours from a) to c), response frequency is confirmed in wide band of a which is vertical axis. In case of Type U, there is no striped pattern from 0.0 to 1.0 in a.

Such this tendency can be also confirmed in case of Type 8. In the case 8, Type 8 doesn't occur so that a wavelet pattern corresponding to Type 8 is explained by using the case 3.

From Figs. 6-e) and 7-b) which results correspond to Type U, in range from 2.0 to 3.0 in a, we can confirm a clear striped pattern. However, the striped pattern is broken around a=1.0 when the orbit gets Type 8.

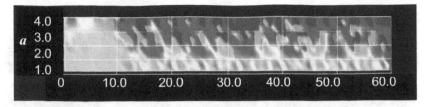

a) T=1.15 s, orbit corresponding to Type *O*

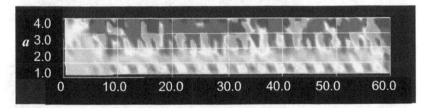

b) T=1.25 s, orbit corresponding to Ordinary Type

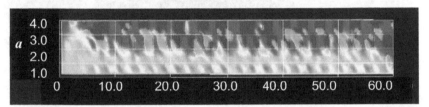

c) T=1.55 s, orbit corresponding to Type *Net* 1

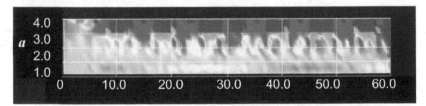

d) T=2.00 s, orbit corresponding to Type *Net* 2

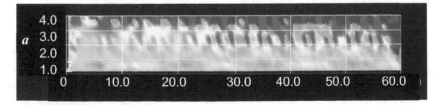

e) T=2.30 s, orbit corresponding to Type *U*

Fig. 6. Patterns of wavelet analysis of Case 8

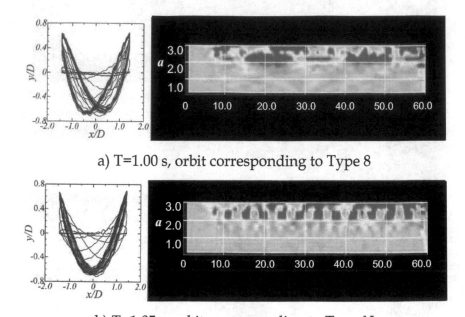

a) T=1.00 s, orbit corresponding to Type 8

b) T=1.05 s, orbit corresponding to Type U

Fig. 7. Patterns of wavelet analysis of Case 3

5.3 Wavelet characteristics of double cylinders in tandem

Results of the vertical bending moment corresponding to VIV with the wavelet analysis are discussed using the results in case of the model 1 in Table 1-b). Correspondence between the orbit and the wavelet pattern was similar to cases of the single cylinder. When the distance ratio increased, VIV behavior resembled the single cases. Therefore the results in case of the distance ratio $s=2.0$ are described here.

Figures 8 to 13 show results of the wavelet analysis, the orbit and a power spectrum of the vertical bending moment. Representations of "front" and "back" in the figures mean follows. The cylinder set on a side of starting direction of the forced oscillation corresponds "front" such as Fig. 3. "f" is a frequency Hz and "f_0" is also a frequency Hz of the forced oscillation on the horizontal axis of the power spectra.

In Figs. 8, the behavior of vibration of both the cylinder is quite different. VIV is not induced strongly. VIV lock-in is confirmed in Figs. 9 to 12. However the behavior of the vibration is different each case from the orbit, even then a shape of the power spectrum of Figs. 9 to 12 are same very much. VIV occurs as bi-harmonic vibration to the frequency of the forced oscillation. The third harmonic vibration can be found on the power spectrum in Figs. 12-a). When the character of "8" is broken and becomes flat, a response component which is higher than the bi-harmonic frequency emerges.

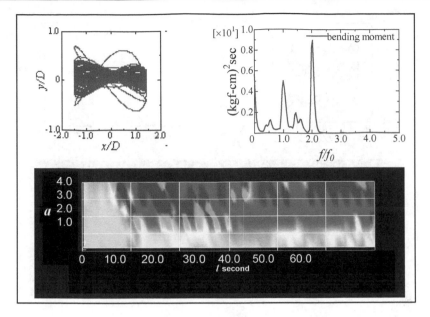

a) on front cylinder

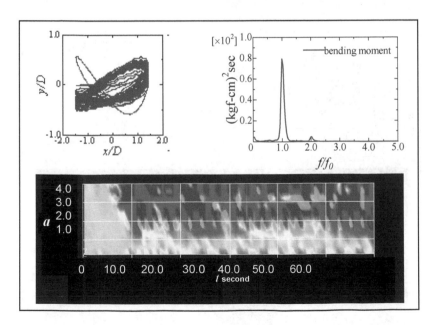

b) on back cylinder

Fig. 8. Comparisons of orbit, power spectrum and wavelet pattern of vertical bending moment in case of s=2 and T=1.2 s

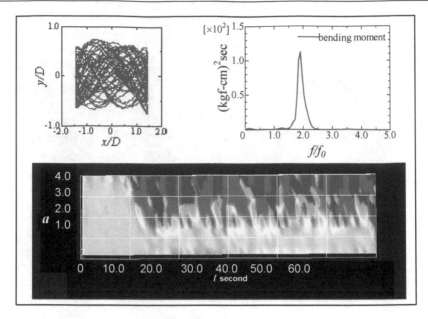

a) on front cylinder

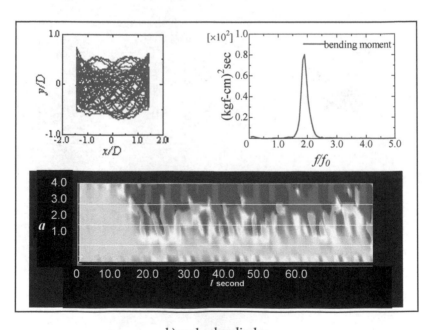

b) on back cylinder

Fig. 9. Comparisons of orbit, power spectrum and wavelet pattern of vertical bending moment in case of s=2 and T=2.0 s

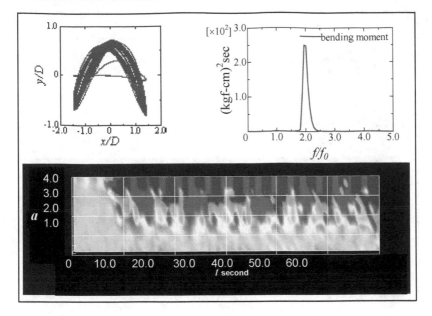

a) on front cylinder

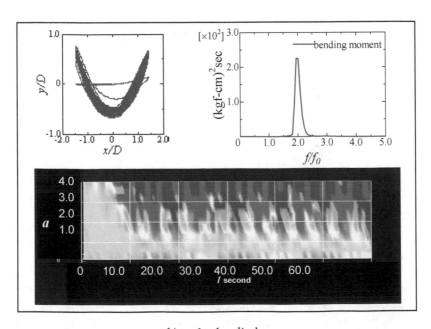

b) on back cylinder

Fig. 10. Comparisons of orbit, power spectrum and wavelet pattern of vertical bending moment in case of $s=2$ and T=2.2 s

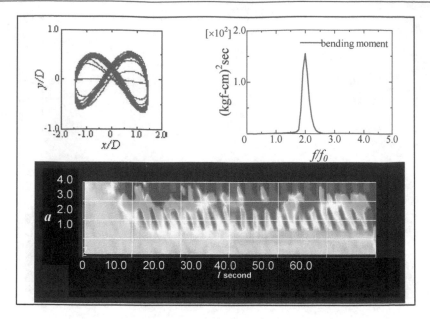

a) on front cylinder

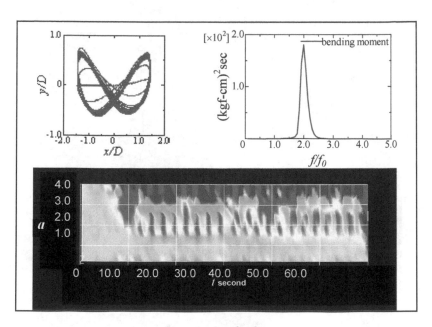

b) on back cylinder

Fig. 11. Comparisons of orbit, power spectrum and wavelet pattern of vertical bending moment in case of s=2 and T=2.4 s

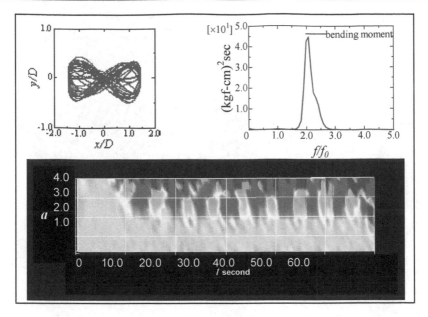

a) on front cylinder

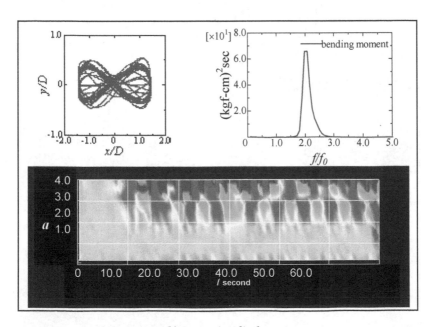

b) on back cylinder

Fig. 12. Comparisons of orbit, power spectrum and wavelet pattern of vertical bending moment in case of s=2 and T=2.8 s

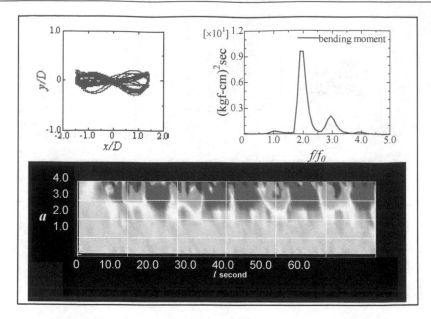

a) on front cylinder

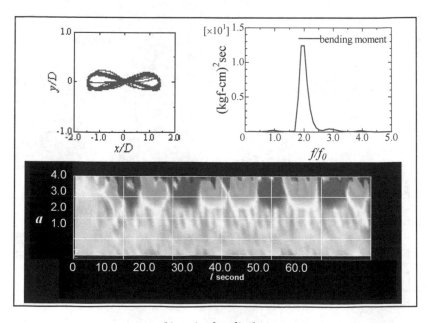

b) on back cylinder

Fig. 13. Comparisons of orbit, power spectrum and wavelet pattern of vertical bending moment in case of $s=2$ and T=3.45 s

Above discussions can be explained from results of the orbits and the power spectra. However all of the wavelet patterns, comparing with results of each period of the forced oscillation, are clearly different. In particular, difference of the VIV behavior cannot be understood from only the power spectra drawing the bi-harmonic vibration such as Figs. 12 to 15. We can found the unique striped pattern on the wavelet contours. At first we notice that the striped pattern gets blurred when the orbit is more complex. When the stripe becomes clearer, in range from 1.5 to 2.5 in *a*, the orbit varies from the *Net* type to the 8 type through the *U* type. From the wavelet pattern, we can know vibration behavior of the cylinder including cross-flow and in-line vibration. In Figs. 16, it can seem that not only the third order but also the fourth order vibration component appear. At around 2.5 in *a*, the wavelet stripes of Fig. 16 get blurrier than that of Fig. 15. From these, it can be considered that a result with the wavelet analysis is higher resolution or more sensitive than that with the FFT analysis to frequency components. Therefore detail of vibration behaviors of cylindrical structures with VIV can be investigated by using the wavelet transform analysis.

6. Conclusion

In this paper, the wavelet transform was applied to the analysis of time histories of vibration of circular cylinders with the vortex induced vibration. From the results, the summary is as follows:

- The orbit pattern of the cylinder roughly corresponds to the unique pattern of the wavelet contour. Therefore the vibration behavior can be known from time history data of arbitral vibration with the wavelet analysis. However calibration is necessary.
- Results with the wavelet analysis are more sensitive than that with the FFT analysis to frequency resolution.
- When VIV lock-in occurs, the pattern of the wavelet contour becomes to clear stripes.
- The Gabor's mother wavelet function is useful for analysis of VIV. In addition, the wavelet transform analysis is effective in order to investigate VIV detail.

7. References

Ikoma, T.; Masuda, K.; Maeda, H. & Hanazawa, S. (2007) *Behaviors of Drag and Inertia Coefficients of Circular Cylinders under Vortex-induced Vibrations with Forced Oscillation Tests in Still Water*, Proceedings of OMAE'07, CD-ROM OMAE2007-29473, ASME

Khalak, A. & Williamson C.H.K. (1999) *Motions, Forces and Mode Transitions in Vortex-induced vibration at low mass-damping*, Journal of Fluids and Structures, Vol.13, pp.813-851

Masuda, K.; Ikoma, T.; Kondo, N. & Maeda, H. (2006) *Forced Oscillation Experiments for VIV of Circular Cylinders and Behaviors of VIV and Lock-in Phenomenon*, Proceedings of OMAE'06, CD-ROM OMAE2006-92073, ASME

Shi, C.; Manuel, L.; Tognarelli, M.A. & Botros, T. (2010) *On the Vortex-Induced Vibration Response of a Model Riser and Lacatin of Sensors for Fatigue Damage Prediction*, Proceedings of OMAE'10, CD-ROM OMAE2010-20991, ASME

Williamson, C.H.K. & Roshko, A. (1988) *Vortex formation in the wake of an oscillating cylinder,* Journal of Fluids and Structures 2, pp.355-381

12

Application of Wavelets Transform in Rotorcraft UAV's Integrated Navigation System

Lei Dai[1,2], Juntong Qi[1], Chong Wu[1,2] and Jianda Han[1]
[1]State Key Laboratory of Robotics,
Shenyang Institute of Automation, Chinese Academy of Sciences
[2]Graduate School of Chinese Academy of Sciences
People's Republic of China

1. Introduction

Rotorcraft UAV (RUAV) has similar mechanical structure with helicopter. It can be operated in different flight modes which the fixed-wing UAV is unable to achieve, such as vertical take-off/landing, hovering, lateral flight, pirouette, and bank-to-turn. For these advantages, RUAV can be used in many fields where human intervention is considered difficult or dangerous (Napolitano et al., 1998). So it can perform the tasks such as regional surveillance, aerial mapping, communications relay, power-line inspection, aerial photography and precision load dropping, etc. RUAV has many advantages, such as small in size, low cost, simple operation and convenient transportation. Therefore, RUAV has broad application prospects, high demands, and advantages that the fixed-wing unmanned aircrafts and unmanned airship can not replace.

Integrated navigation system can give the movement information of the carrier, thus every UAV has an integrated navigation system. Because of the limitations of weight, volume, power supply and cost, there is no redundant navigation system in RUAV. RUAV does not have the emergency landing properties of fixed-wing aircrafts or airships in case of failures. Therefore, a failure in any part of a RUAV can be catastrophic. If the failure is not detected, identified and accommodated, the RUAV may crash. The use of wavelet transforms the situation of accurately localizing the characteristics of a signal both in the time and frequency domains, the occurring instants of abnormal status of a sensor in the output signal can be identified by the multi-scale representation of the signal (Dabechies, 1988; Isermann, 1984; Zhang, 2000). Once the instants are detected, the distribution differences of the signal energy on all decomposed wavelet scales of the signal before and after the instants are used to claim and classify the sensor faults.

In low cost and small size integrated navigation system, MEMS (Micro Electronic Mechanical System) inertial sensors are used widely. But MEMS inertial sensors, especially MEMS gyroscopes have large noise. It affects the calculation accuracy of angle rotation matrix, and will further affect calculation accuracy of other navigation data such as position, velocity, and angular velocity. In order to improve the calculation precision of position and angle, digital filter is required to reduce the noise of gyroscope. Commonly, we used

Kalman filters to decreasing the random noise. And we need to build the mathematic model of sensors' errors. The MEMS gyroscope has random drift characteristics of weak nonlinear, non-stationary, slow time-varying. And it is sensitive to external environments such as vibration and temperature. The result of Kalman filter is often imprecision and even divergence, because of inaccurate drift error model of MEMS gyroscope. Wavelet transform has the characteristics of multi-resolution and time-frequency localization, and we do not need to build the mathematic model of sensor errors. So it is ideal for signal processing and analysis of MEMS gyroscope.

However, Synthetic data simulated by means of a computer using real flight data from ServoHeli-20 and ServoHeli-40 RUAV, which is designed and implemented by Shenyang Institute of Automation, have verified the effectiveness of the proposed method.

The following part of this paper is organized as follows. In Section 2, the fault detection approach based on the wavelet transform is established. The RUAV verification platform is introduced in Section 3. The integrated navigation system and the characteristic of inertial sensor are discussed in Section 4. Real RUAV flight fault detection experiments in manual mode are described and discussed in Section 5, and conclusions are given in Section 6.

2. The RUAV platform

During years of research, we have developed two RUAV platforms. The ServoHeli-20 RUAV platform was designed to be a common experimental platform for control and fault-tolerant related study before 2009 (Qi, et al., 2006). The hardware components were selected with considerations of weight, availability and performance. After that, we miniaturized the hardware and used 40Kg industry helicopter. And we developed ServoHeli-40 RUAV platform. This RUAV platform can be used for control and navigation algorithm verification and experimental payload platform.

2.1 ServoHeli-20 platform

As the basic aircraft of the RUAV system, we chose the small-scaled model helicopter which is available in the market. Such a choice is easy for us to exchange the accessories and reduce the cost (Qi, et al., 2010).

2.1.1 Modified RC helicopter

ServoHeli-20 aerial vehicle is a high quality helicopter which is changed by us using a RC model helicopter operating with a remote controller. The modified system allows the payload of more than 5 kilograms, which is sufficient to carry the whole airborne avionics box and the communication units. The fuselage of the helicopter is constructed with sturdy ABS composite body and the main rotor blades are replaced by heavy-duty carbon fibre reinforced ones to accommodate extra payloads. The vehicle is powered by a 90-class glow plug engine which generates 3.0hp at about 15000 rpm, and practical angular rate ranging from 2,000 to 16,000 rpm. The full length of the fuselage is 1260mm and the full width of it is 160mm. The total height of the helicopter is 410mm, the main rotor is 1600mm and the tail rotor is 260mm.

The overall rotorcraft UAV control system comprises: the aerial vehicle platform, the onboard avionics control system, and the ground monitoring station. The UAV helicopter itself is able to operate with the independent control computer system and onboard sensors.

2.1.2 Navigation sensors

In order to navigate following a desired trajectory while stabilizing the vehicle, the information of helicopter's position, velocity, acceleration, attitude, and the angular rates should be known to the guidance and control system. The ServoHeli-20 RUAV system is equipped with sensors including IMU (Inertial Measure Unit), GPS and digital compass, to obtain above accurate information about the motion of the helicopter in association with environment.

Sensor	PARAMETERS
IMU400 IMU	Angular Rate Range: $\pm100°$/sec Acceleration Range: $\pm4g$ Digital Output Format: RS-232 Update Rate: > 100 Hz Size: 76.2 × 95.3 × 81.3 mm Weight: <640g
Crescent GPS	Position Accuracy (CEP): 1.5m Digital Output Format: RS-232 Update Rate: 10 Hz Size: 71.1 × 49.6 × 1.2 mm Weight: 20g
HMR3000 Compass	Pitch, Roll Angular Range: $\pm40°$ Digital Output Format: RS-232 Update Rate: 20 Hz Size: 15 × 42 × 8.8 mm Weight: 92g

Table 1. Sensors parameters

The picture of sensors in the avionics box is shown in Fig. 1, and their primary parameters are shown in Table I.

2.1.3 Processor and control system

The flight computer installed in avionics box is a typical industrial embedded computer system, so-called PC-104 in the whole system is kept as compact and light-weight as possible. The PC-104 has the ISA or PCI bus which features a 108.2cm × 115.06cm footprint circuit board. Our flight computer system consists of a main CPU board and some other peripheral boards such as DC-DC power supply board, 8-channel serial communication device and PWM generation board. The main CPU board has a Celeron processor at 400MHz with 256MB SDRAM, fully compatible with the real-time operation system such as QNX. Hard drive or other equivalent mass-storage device for booting and running an operation system and storing useful sensor data is needed to the flight computer.

To our flight control system, a real-time operation system (RTOS) is required for the onboard computer system. After carefully consideration and comparison, QNX Neutrino RTOS is selected as the operation system, which is ideal for embedded real-time applications. It can be scaled to very small size and provide multitasking threads, priority-driven pre-emptive scheduling, and fast context-switching–all essential ingredients of an embedded real-time system. The applied program can be coded and debugged in the remote windows-host computers and can be executed in the airborne computer system independently, which provides great convenience during the flight experiments without modifying the program in onboard computer.

2.1.4 Implementation

Designing the avionics box and packing the box appropriately under the fuselage of the helicopter are two main tasks to implement of the RUAV system.

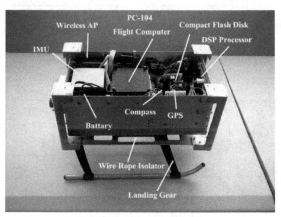

Fig. 1. The avionics control system

In the actual flight environment, the weight and the size of the avionics box are strict limited. Our airborne control box, which is shown in Figure 1, is a compact aluminum alloy package mounted on the landing gear. The center of gravity of the box lies on the IMU device where is not the geometry center of the system that ensure the navigation data form IMU accurate. The digital compass and the IMU are installed on the same line, which are taken as the horizontal center of the gravity of the avionics system to locate and the other components.

The original landing gear of the model helicopter is plastic, in which is no enough room to install the designed avionics system in the fuselage of the helicopter. While, we re-design a landing gear with aluminum alloy and make a larger room under the fuselage of the model helicopter for the control box. To avoid the disciplinary vibration about 20Hz caused by characteristic of the helicopter, ENIDINE® aviation wire rope isolators which are mounted between the avionics box and the changed landing gear are chosen. They are comprised of stainless steel stranded cable, threaded through aluminum alloy retaining bars, crimped and mounted for effective vibration isolation. The assembled RUAV system with the necessary components is shown in Figure 2.

Fig. 2. Implemented ServoHeli-20 RUAV

The full duplex wireless-LAN equipments are installed in the ground station and the airborne system to exchange data between them including receiving commands from the ground system and reporting the operating status or possible damages to the ground station. The architecture of the RUAV control system is presented in Figure 3.

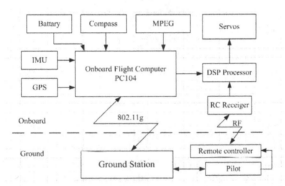

Fig. 3. Architecture of the RUAV control system

2.2 ServoHeli-40 platform

ServoHeli-40 RUAV platform is a flexible platform. It can carry 15Kg payloads to do aerophotography or experiment. It is also easy for us to replace the accessories and the cost is affordable.

2.2.1 ServoHeli-40 helicopter

ServoHeli-40 aerial vehicle is a high reliability helicopter. It uses traditional helicopter configuration of single-rotor with tail rotor. The power is come from 8hp twin-cylinder two-stroke air-cooled gasoline engine. The fuselage of the helicopter is constructed with aluminium. The bearings and the other parts are standard industry parts. So it is easy to buy and repair. ServoHeli-40's maximum takeoff weight is 40Kg, and its maximum payload is 15 kg. It can carry different kinds of instruments to do experiment. The rotor diameter is

2150 mm. The total height of the helicopter is 770mm, the full width of it is 720mm, and the total length is 2680mm. The maximum airspeed is 100 km/h, and it can cruise 1 hour at 36 km/h speed.

2.2.2 Navigation system

In the design of the navigation system, we followed some principles. First, the system must be compact and easy to equip on the airframe. Second, the system should use low-cost sensor to reduce the RUAV system's cost. Third, the system should be designed as light as possible to save fuel and increase the payload. Precision navigation information of flight state is needed to realize the autonomous control of the RUAV. Generally, navigation information must include positions, velocities, accelerations, attitude, heading and angular velocities in 3-axis. The architecture of the navigation system is shown in Figure 4 (Wu, et al., 2010).

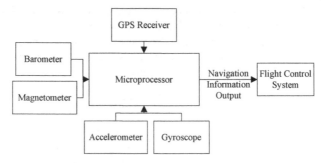

Fig. 4. Hardware Architecture of Navigation System

We use GPS receiver to give position and velocity. But the altitude given by GPS generally has a fluctuation of 5m, and it is not accurate enough to control the RUAV. So a barometer is needed to give a more accurate measurement of the relative altitude. Even though the barometer is more accurate in relative altitude, it is susceptible to weather condition and may vary significantly in different weather. Because the altitude given by GPS is much more unsusceptible to weather, a combination of GPS altitude and barometer altitude will give more accurate and stable altitude information.

Attitudes' accuracy is the key point for the stability of the RUAV since the position control of the RUAV is coupled with the attitude. We use IMU to measure the acceleration and angle velocity. By referring to a low-cost attitude design described in literature (Gao et al., 2006), we determine the attitude in pitch and roll by accelerometers and gyroscopes. A simple calculation of acceleration may be suitable for the RUAV in hovering and other mode with low maneuverability in low acceleration, but in high maneuverability mode with high acceleration, the measurement will deviate a great deal because the measurement of accelerometers include not only gravity acceleration but also absolute acceleration. So a feedback from velocity is added to decrease the influence of absolute acceleration.

The yaw of RUAV will be calculated by the magnetic field measured by the magnetometer. Due to the deviation of magnetic field of the earth in different places, a revision of magnetic field would be necessary to get the real yaw.

Compared with previous generation, this navigation system uses independent processor. We choose LPC3250 as the calculation and acquisition processor. A two-stage EKF is implemented to estimate the flight state. Further research in navigation theory can be conducted by using this system.

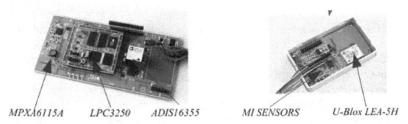

MPXA6115A   LPC3250   ADIS16355   MI SENSORS   U-Blox LEA-5H

Fig. 5. The navigation system

The IMU, GPS, barometer, magnetometer and processor are integrated in a compact circuit, as shown in Figure 5. The primary parameters of compacted navigation system are shown in Table2.

Sensor	PARAMETERS
ADIS16355 IMU	Angular Rate Range: ±300° / sec Acceleration Range: ±18g Digital Output Format: SPI Update Rate: 50 Hz Size: 23 × 23 ×23mm
U-Blox LEA-5H GPS	Position Accuracy (CEP): 3m Digital Output Format: Uart Update Rate: 4Hz Size: 17 × 22.4 × 3mm
SmartSens Magneto-inductive Compass	Pitch, Roll Angular Range: ±40° Analog Output Update Rate: 20 Hz
MPXA6115A Barometer	Sensor range: 15 to 115 kPa Analog Output Update Rate: 50 Hz Size: 7.5 × 10 ×10mm

Table 2. Sensors Parameters

2.2.3 Flight control system

The fight control processor uses the same processor as the navigation system, so we can share the code of the operating system and reduce the debug time. The LPC3250 with an ARM926EJ-S CPU Core implementation uses Harvard architecture with a 5-stage pipeline and operates at CPU frequencies up to 266 Hz. The Vector Floating Point (VFP) coprocessor makes the micro controller suitable for advanced navigation and control algorithm, and processing speed and interface versatility is guaranteed. The industry standard operation temperature from -40℃ to 80℃ extends the usage of RUAV in various environments. The LPC3250 includes a USB 2.0 Full Speed interface, seven UARTs, two I²C interfaces, two SPI/SSP ports, and two I2S interfaces; Such a great number interfaces of LPC3250 makes it very suitable for navigation and control system with a plenty of sensors in standard interface. We designed interface circuit to drive the actuator and log the flight data.

To decrease the developing work in programming, while increasing the system stability, a μC/OS-II embedded system is installed to organize the software development. This small sized embedded system is quite convenient to install; the hard-real-time architecture also makes it suitable for a time critical avionics system in RUAV. We divided the work of software into 5 parts. First, the OS Kernel is to maintain the whole system and arrange the task schedule. Second, the algorithms implements navigation and control theory. Third, the device interface process is to handle the task for sensor data acquire and drive the actuator. Fourth, the user interface carries out the job to display and receive necessary information to the user. Fifth, the log interface is to log the flight data for our experiment. To make sure that the algorithms can be calculated in time, a hardware timer is used instead of the software timer provided by operator system. With a proper design of the software architecture, the system's stability is maintained and the flexibility is also provided for other algorithm implementations.

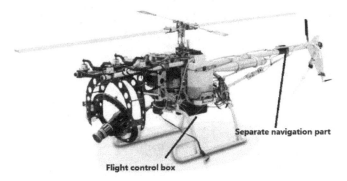

Fig. 6. Implemented ServoHeli-40 RUAV

2.2.4 System realization

The GPS receiver and magnetometer are in a separate part and the others are in the main navigation part. The flight control system and main navigation system are assembled in an anti-jamming aluminum box, and called flight control box. Such a separation is with the consideration that the GPS and magnetometers are susceptible to the install position because they may be influenced if it is covered by the airborne or near some magnetic material. The flight control box is mounted under fuselage of the RUAV. The separate part can be equipped in a proper place on the airframe. To avoid the disciplinary vibration about 20-22.5Hz caused by revolving of main rotor, ENIDINE aviation wire rope isolators are also used. They are comprised of stainless steel stranded cable, threaded through aluminum alloy retaining bars, crimped and mounted for effective vibration isolation. The assembled RUAV system with the necessary components is shown in Figure 6.

To increase the control distance and reliability, the half duplex industry radios are installed in the ground station and the airborne system to exchange data. The data includes commands, operating status and possible damages, which is received and reported to the ground station. The flight data can be logged in SD card, we can analysis the data after flight test. The architecture of the RUAV control system is presented in Figure 7.

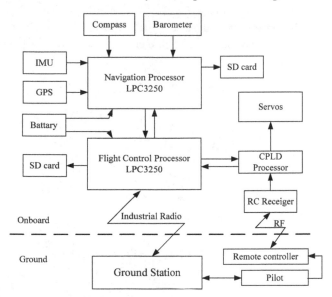

Fig. 7. Architecture of the RUAV control system

3. Wavelet-based fault detection

Without loss of generality, assume that the vehicle's sensor output $y(t)$ is described as (Zhang & Yan, 2001):

$$y(t) = f[x(t)] + n(t) \tag{1}$$

Where $n(t)$ is a noise signal and the measured $x(t)$ changes with a k-degree polynomial function $f[x(t)]$ which describes the measured process changes. Stone-Weierstrass Theorem states that any continuous function on a compact set can be approximated to any degree of accuracy by a polynomial function (Rudin, 1976). Therefore, using a polynomial function to represent any function $f[x(t)]$ will not lose the generality. Let $\psi(t)$ be a wavelet function and $\psi_s(t) = (1/s)\psi(t/s)$ be the dilation of $\psi(t)$ by the scale factor s. The wavelet transform of $y(t)$ can be written as:

$$WT_f(s,\tau) = y(t) * \psi_s(t) = f[x(t) + n(t)] * \psi_s(t) \tag{2}$$

Where $*$ denotes the convolution and $WT_f(s,\tau)$ represents the wavelet transform. A wavelet $\psi(t)$ is said to have m vanishing moments if and only for all positive integers $k<m$, the following equation is satisfied:

$$\int_{-\infty}^{+\infty} t^k \psi_s(t)dt = 0 \tag{3}$$

Now, let us call a smoothing function, any real function $\theta(t)$ such that $\theta(t) = O(1/1(1+t^2))$ and whose integral is nonzero. A smoothing function can be viewed as the impulse response of a low-pass filter. Let $f[x(t)]$ and $\theta_s(t) = (1/s)\theta(t/s)$ be a real function in $L^2(R)$. The abrupt changes of the sensor data at scale s are defined as local sharp variation points $f[x(t)]$ smoothed by $\theta_s(t)$. The method of detecting these sharp variation points with a wavelet transform is explained as follows.

Let $\psi^1(t)$ and $\psi^2(t)$ be the two wavelets defined by:

$$\psi^1(t) = \frac{d\theta(t)}{dt} \tag{4}$$

$$\psi^2(t) = \frac{d\theta^2(t)}{dt^2} \tag{5}$$

The wavelet transform defined with respect to each of these wavelets are given by:

$$W^1(t) = f * \psi_s^1(t) = f * (s\frac{d\theta_s}{dt})(t) = s\frac{d}{dt}(f * \theta_s)(t) \tag{6}$$

$$W^1(t) = f * \psi_s^1(t) = f * (s\frac{d\theta_s}{dt})(t) = s\frac{d}{dt}(f * \theta_s)(t) \tag{7}$$

The wavelet transforms of $W^1f(s,t)$ and $W^2f(s,t)$ is proportional respectively to the first and second derivatives of $f[x(t)]$ smoothed by $\theta_s(t)$. As a result, the local maxima of $|W^1f(s,t)|$ indicate the locations of sharp variation points and singularities of $f[x(t)] * \theta_s(t)$ (Mallat & Hwang, 1992).

From (4) to (7), it can be concluded that the wavelet transform of the signal (1) only includes some sharp variation points induced by sensor faults and random noise. Once a sharp variation point is claimed, and alarm will be triggered for a failure of the sensor.

4. Fault detection experiment

4.1 Fault detection system design

The sensors of the navigation system with different mechanism also have different performance. We cannot get the ideal fault detection results using the traditional fault detection techniques.

In order to accompany the short control period and the highly update rate, we use the parallel wavelet analyzer, which is shown as figure 8.

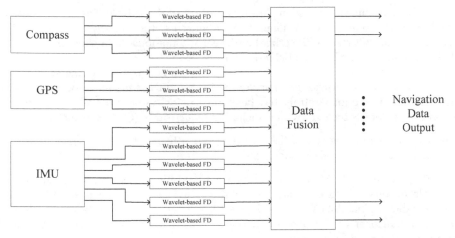

Fig. 8. Architecture of wavelet-based sensor system

Table 1 shows us that there are three sensors in the RUAV sensor system which have three channel separately. We design 12 wavelet analyzers for 12 channels of all sensors. The sensors data will directly send to the data fusion system when the data are in the normal states. However, if the sensors data is abnormal in one or some channels as a result of the failure of specific sensors, the alarms will send to the flight computer while the data link will be cut off. Then the navigation system will continue to compute with degraded sensors data.

4.2 Experimental results and discussion

The proposed wavelet-based fault detection system tested using the ServoHeli-20 RUAV system in manual mode.

Fig. 9. ServoHeli-20 fault detection experiment

The use of the mathematical model makes it easier to test the wavelet-based fault detection system, but the characteristic of the datasets may not reflect the real flight environment and the actual actuator failures. On the other hand, real autonomous flight experiments with an injected sensor failure can be potentially dangerous for the helicopter because it can take the RUAV out of control and RUAV may crash. Thus, we planned to inject the sensor failure while the absence of the security problems of the RUAV with its manual mode. As is shown in the figure 9, the pilot controls the helicopter using radio controller. The onboard computer online detects the fault with wavelet-based algorithm (Qi & Han, 2007).

To demonstrate the effectiveness of the fault detection scheme, the failure scenario of abrupt bias and spike in compass roll channel is assumed.

A "db2" ("db" is define in Matlab) wavelet with a vanishing moment 2 is applied to these abrupt faults of sensor. Figure 10 and 11 show their wavelet transforms in scale-D1 to scale-S3 including the original data signals. In figure 10, scale-D1 to scale-D3 denote the details of the wavelet transform of the signals on scales 1 to 3, respectively, while the scale-S3 represents the approximation of them on scale 3.

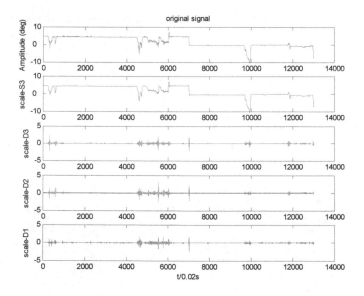

Fig. 10. Bias failure and its wavelet transform(t=0.02s)

4.2.1 Failure of bias

In figure 10, an example sensor failure experiment is presented. At the point of 7000, the compass roll channel gets bias of 5 degree.

The local maxima of the first derivative are sharp variation points of $f[x(t)] * \theta_s(t)$. For abrupt failure detection, we are only interested in the local maxima of $|W^1 f(s,t)|$. When detecting the local maxima of $|W^1 f(s,t)|$, we call also keep the value of the wavelet transform at the corresponding location.

As is shown in the figure 11, discontinuity point of signal is displayed obviously, it is allocated very accurately in time-domain, and fault point of bias signal is contained in signal abruption. Using the $|W^1 f(s,t)|$ criterion, the fault detection system can detect locations of the bias fault at 7000 that we can see the local maximum value of module indicates the signal singularity accurately.

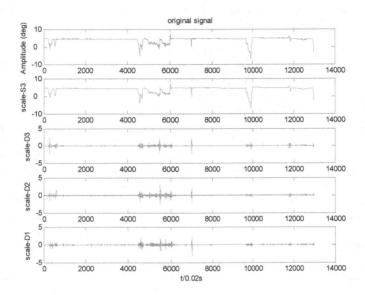

Fig. 11. Spike failure and its wavelet transform(t=0.02s)

4.2.2 Failure of spike

We also made a spike failure injection to RUAV system in manual mode to test the performance of the wavelet-based fault detection system. At the point of 7000, the compass roll channel gets spike which the signal return to zero.

Similar to the bias failure experiment, the location of fault agree with the maximum values of the wavelet transform on different scales.

From the results, it can be conclude that the proposed method is effective for detection the abrupt faults of the RUAV sensor system. Fault point could be also being described accurately at some certain resolution. Local characteristics of wavelet are represented well in time and frequency-domains.

5. Wavelet for gyroscope de-noising

5.1 MEMS gyroscope signal analysis

With the development of microelectronics technology, low cost MEMS gyroscopes begin to be used widely. It makes the great development of integrated navigation system, especially in UAV system. Compared with high costs gyroscope, the MEMS gyroscope devices have some drawbacks, such as large bias stability, big temperature noise, high noise density. And all these disadvantages lead to that the long-term accuracy of navigation system is very low. While random vibration due to main rotor of RUAV, have an impact on gyroscopes measurements. Simple passive vibration damping measures cannot be completely filtering the vibration. And the measurement error is unacceptable (Ma et al., 2007). In order to eliminate noise of MEMS gyroscope, error analysis of signals need to be done. This is very important to improve the performance of integrated navigation system and increase the stability of the RUAV system.

According to the frequency spectrum characteristics of MEMS gyroscope, its errors can be divided into long-term errors and short-term errors. Long-term errors include bias stability, scale non-linearity, angular random walk, bias variation over temperature, rate noise density, and so on. These errors can be predicted by the mathematic model and adjusted. According to the error mechanism of MEMS gyroscope, an ARMA model is established. Then, using parameter identification, the parameter of the ARMA model is identified. So the long-term errors can be compensated. The short-term errors include random interference noise, measurement noise, and so on. It is a tricky problem to deal with errors. Usually, we use digital filters to compensate the error. These conventional denoising methods include Low-Pass filter, Kalman filter, and wavelet filter. Under the principle of linear least mean squares error, angular velocity estimation is recursively calculated by Kalman filtering in literature (Shi & Zhang, 2000). Although the approach is successfully used in reducing gyroscope noise on the stationary platforms, it is based on the assumption that the signal is corrupted by Gaussian noise and model is exactitude. Unfortunately, for imprecise model and colored noise, this method may yield worse results. Low-Pass filter passes low-frequency signals but reduces the amplitude of signals with frequencies higher than the cutoff frequency. And greater accuracy in approximation requires a longer delay. It can be realized by cheap hardware, while its low quality, however, is not very satisfactory. Wavelet transforms have excellent multi-resolution analysis feature and do not need model. So it is suitable for non-stationary signals processing. And wavelet transforms have been successfully applied to the denoising of signals or images in recent years. This method has achieved good results in gyroscope signal denoising process (Imola et al., 2001; Qu et al., 2009).

5.2 Wavelet for denoising

Wavelet transform is method of time-frequency localization analysis. Its window size (area) is fixed, but the shape can be changed. Wavelet transform developed the short - time Fourier transform of localized. It has a high frequency resolution and lower time resolution in low frequency part of the signal. And in high frequency part of the signal, it has a high time resolution and lower frequency resolution. Wavelet transform method has the character of frequency analysis, and it can indicate the occurred time. It is very suitable for noise reduction of MEMS gyroscopes.

The discrete wavelet function $\psi_{j,k}(t)$ in Discrete Wavelet transform (DWT) can be expressed as :

$$\psi_{j,k}(t) = s_0^{-j/2}\psi\left(\frac{t - ks_0^j\tau_0}{s_0^j}\right) = s_0^{-j/2}\psi\left(s_0^{-j}t - k\tau_0\right) \tag{8}$$

The discrete wavelet translate factor can be expressed as :

$$C_{j,k} = \int_{-\infty}^{\infty} f(t)\psi^{*}_{j,k}(t)dt = \langle f, \psi_{j,k} \rangle \tag{9}$$

The reconstruction function of Discrete Wavelet transform can be expressed as :

$$f(t) = C\sum_{-\infty}^{\infty}\sum_{-\infty}^{\infty}C_{j,k}\psi_{j,k}(t) \tag{10}$$

Where s denotes the scale factor, τ denotes the translate factor. The $f(t)$ represents the signal function.

The purpose of wavelet transform for denoising is to extract useful signal and remove the interference signal in the output signal. In other words, the useful signal and noise signal are separated by the method of wavelet transform. There are 4 common methods of wavelet denoising (Burrus et al., 1998; Guo et al., 2003):

1. Thresholding Denoising Method. It is also called wavelet shrinkage. The basic idea of this method can be described as: The wavelet coefficients have different characters in particular wavelet scales. According to this characteristic of the signal and noise, the noise signals are converted by wavelet transform in certain wavelet scales. According to a certain threshold processing strategies for treatment of wavelet coefficients, the coefficients, which are greater than the threshold, are kept (hard thresholding method) or shrunk (soft thresholding method). The coefficients, which are less than the threshold, are considered to be noise and set to zero directly. Then based on these wavelet coefficients, the original signal is reconstructed using inverse wavelet transform. And this method requires the assumption that the noise signal is Gaussian white noise.
2. Wavelet Decomposition and Reconstruction Method. It is also known as the Mallet method. It decomposes the signal with noise in scale into different frequency bands, sets the bands with noise to zero and reconstructs the signal using wavelet method. This method will remove the noise signals with the useful signals. So it may distort the reconstructed signal.
3. Modulus Maximum Method. In different wavelet scale, this method uses the variation features of wavelet transform modulus' maxima value to denoising the signal. The extreme points, whose amplitude decrease with scale increasing in signal, are removed. The extreme points, whose amplitude increase with scale increasing in signal, are retained. Using alternating projection method, the original signal is reconstructed from de-noised diagram of maxima modulus. And the noise signal is de-noised.
4. Translation Invariant Method. It is a method improved from the basis of the Thresholding Denoising method. The noise signals are taken n times cycles shift by this method. And the translated signals are de-noised using thresholding denoising method. In the end, de-noised signal are equilibrated. This method has a smaller mean square error and improves signal-to-noise ratio.

5.3 Thresholding denoising method

Through the above analysis, the modulus maximum method and translation invariant method have large calculation amounts. And this will affect the real-time calculation of integrated navigation system. So considering the speed of calculation and the ease of implementation, thresholding denoising method is used in our navigation system.

The step of thresholding denoising method is as follows (Song et al., 2009; Su & Zhou, 2009):

1. Selecting the Wavelet function. Then the signal with noise $y_i, i = 0,1,...,N-1$ is discrete using wavelet transformation. A group of wavelet transform coefficients $d_{j,k}$ is got. The subscript j is the wavelet scale.
2. Thresholding the wavelet transforms coefficients $d_{j,k}$. The hard threshold, soft threshold or other threshold method can be used to deal with the coefficients. After the computation, a new wavelet transforms coefficients $\hat{d}_{j,k}$ is got.

The hard threshold estimation is defined as follows:

$$\hat{d}_{j,k} = \begin{cases} d_{j,k}, & |d_{j,k}| \geq \lambda_j \\ 0, & |d_{j,k}| < \lambda_j \end{cases} \tag{11}$$

The soft threshold estimation is defined as follows:

$$\hat{d}_{j,k} = \begin{cases} \text{sgn}(d_{j,k})(|d_{j,k}| - \lambda_j), & |d_{j,k}| \geq \lambda_j \\ 0, & |d_{j,k}| < \lambda_j \end{cases} \tag{12}$$

Where λ_j is the threshold constant.

3. Wavelet reconstruction. Using the inverse of discrete wavelet transform formulas, we can get the de-noised signal $\hat{y}_i$.

6. Gyroscope denoising simulation and analysis

In this section, several control experiments are taken on gyroscope signal denoising using wavelet methods. The experiment uses the real flight data recorded by the flight control system of ServoHeli-40. The flight data is recorded at the 100Hz rate, and each point represents 10ms. It is sufficient to describe the motion of ServoHeli-40 RUAV both in time domain and in frequency domain.

In this simulation experiment, the length of the data is 14,950 points. And it means the data continued about 2.5 minutes. During this period of time, the RUAV did the following actions: standing still, engine ignition, speed idle, hovering , and trajectory tracking. In these flying modes, the data of gyroscope have different amplitude characters. This reflects the vibration differences in different flying modes. The original signal of ServoHeli-40's Y axis gyroscope is shown in the figure 12.

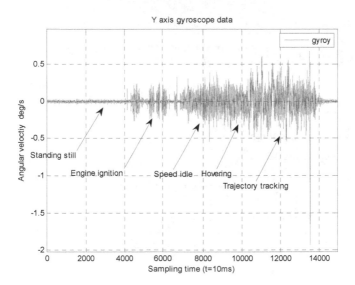

Fig. 12. The original signal of ServoHeli-40's Y axis gyroscope

In order to find a appropriate wavelet functions and decomposition levels, the simulation compared the thresholding denoising method using harr, db2, db4, db6, sym2, sym4, coif2, bior1.5 and bior5.5 wavelet functions. The decomposition levels are respectively 2, 5 and 8. The standard deviation of de-noised signal's residuals is calculated to compare the wavelet denoising results. The results are shown in Table 3. And the standard deviation of the original data is 0.06606.

Level \ wavelet	haar	db2	db4	db6	sym2	sym4	coif2	bior1.5	bior5.5
2	0.01841	0.1378	0.01323	0.01285	0.01378	0.01329	0.01288	0.01906	0.01271
5	0.04229	0.05397	0.05467	0.05489	0.05397	0.05561	0.05558	0.05514	0.05572
8	0.06552	0.06537	0.06576	0.0658	0.06573	0.06578	0.06582	0.06573	0.06583

Table 3. Standard deviation of de-noised signal's residuals

In Table 3, when decomposition level increases to more than 5 layers, improvement in de-noised signal's residuals is unobvious. When the de-noised signal's residuals approach to 0.06606, the de-noised signal is close to straight line. And the computation cost is increased as layers increasing. So decomposition level of 5 is a good choice.

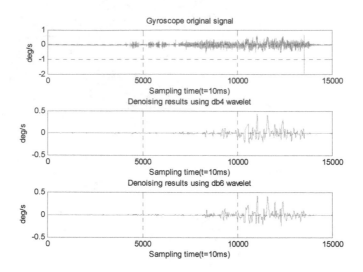

Fig. 13. Contrast of denoised signal and original signal

According to the simulation results, db4, db6 and bior5.5 may be good choice for wavelet functions, because the curve of these de-noised signal are smoother than the others. But bior5.5 have lager computation cost, it is not suitable for real time computation. In Figure 13, the de-noised signals of db4, db6 are compared with the original signal. The denoising result, got from db6 wavelet function, is smoother than the result of db4. And the de-noised signal of db6 is closer to the real angular moment of RUAV than the original signal.

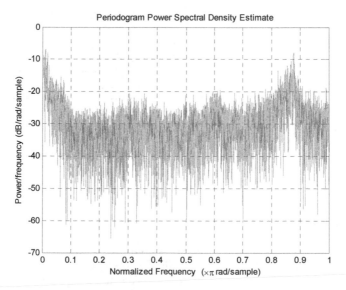

Fig. 14. The periodogram power spectral density estimate of original signal

The periodogram power spectral density estimate of original gyroscope signal is shown in Figure 14. The normalized frequency is 50Hz in this diagram. At 0.87(about 37Hz), the signal has a gain of -10db. In ServoHeli-40 RUAV system, there is a 16.7-45Hz vibration band. This vibration is caused by the rotation of main rotor, engine and tail rotor. 37Hz signal is in this noise band, and it is need to be eliminated.

For the characters of the control system, actuator system, and airframe of helicopter, the motion response of ServoHeli-40 is no more than 3Hz. The vibration frequency larger than 3Hz is out of the control of flight control system. Using the denoising result of db6 wavelet function, the spectrum energy density is analyzed. The periodogram power spectral density estimate of de-noised signal is shown in Figure 15. In this diagram, the noise signal more than 2.5Hz are eliminated by the algorithm. The wavelet filter can just remove high frequency noise. The denoising results reflected the actual movement of the aircraft. This method is suitable for denoising the noise of gyroscope.

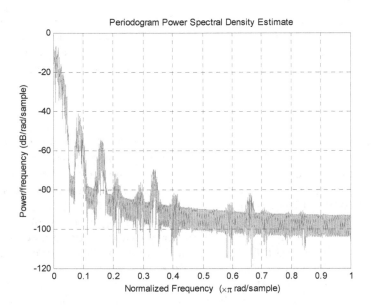

Fig. 15. The periodogram power spectral density estimate of denoised signal using db6 wavelet function

7. Conclusion

In this chapter, wavelet-based algorithm is applied to fault diagnosis and gyroscope noise reduction. Its advantage is that it does not require a prior model of a sensor. The proposed wavelet-based algorithm for fault detection of the RUAV sensor system gives us a multi-scale analysis approach to identify the feature of flight data failures, which are not readily identified by traditional approaches. The results presented in this chapter have shown that the method based on wavelet transform is a promising alternative to other approaches to the fault detection system for RUAV system. With the wavelet-based scheme, the RUAV sensor fault detection system can detect the failure locations of abrupt signal effectively. In order to overcome the drawbacks of the low-pass filter, the thresholding denoising method base on wavelet transformation is used to reduce the short-term measured noise of the MEMS gyroscope. The article compared different wavelet functions and level of decompositions, and found the effective filter parameters. Using db6 wavelet function at level 5, the denoised signal is suitable for integrated navigation system and flight control system. This will improve the calculation precision of angle rotation matrix, and the high frequency noise will be decrease.

In the Further, further flight tests are needed to verify the actual performance of wavelet-based denoising method and wavelet-based fault detection in ServoHeli-40's integrated navigation system.

8. Acknowledgments

This work was partially supported by the National Natural Science Foundation of China, "A novel method about the flight control for flying-robot, based on the characteristics of human's brain decision" (No.61005086). And the authors gratefully acknowledge the contribution of Shenyang Institute of Automation, Chinese Academy of Sciences and reviewers' comments.

9. References

Burrus C. Sidney, Gopinath Ramesh A, Guo Haitao. (1998). *Introduction to Wavelets and Wavelet Transforms: A Primer*. Prentice Hall, Inc, 0134896009, New Jersey.

Daubechies Ingrid. (1988). Orthonormal bases of compactly supported wavelets. *Communications on Pure Applied Mathematics*, Vol.41, No.7, (October 1988), pp.(909–996), 10.1002/cpa.3160410705.

Gao Tongyue, Gong Zhenbang, Luo Jun, Ding wei, Feng wei. (2006). An Attitude Determination System For A Small Unmanned Helicopter Using Low-Cost Sensors. *Proceedings of the 2006 IEEE International Conference on Robotics and Biomimetics*, 1-4244-0570-X, Kuming China, December 2006.

Guo Jiang-chang, Teng Jian-fu, Zhang Ya-qi. (2003). The denosing of gyro signals by bi-orthogonal wavelet transform. *IEEE Canadian Conference on Electrical and Computer Engineering*, 0-7803-7781-8, Montreal, Canadian, May 2003.

Imola. K. Fodor, Chandrika Kamath, Rika Kamath. (2001). Denoising through wavelet shrinkage: An empirical study. *J. Electron. Imaging*, Vol. 12, No.1,(2003), pp. (151 – 160), 10.1117/1.1525793.

Isermann Rolf. (1984). Process fault detection based on modeling and estimation methods—A survey. *Automatica*, Vol.20, No.4, (1984), pp. (387–404), 10.1016/0005-1098(84)90098-0

Ma Jianjun, Zheng Zhiqiang, Wu Mei-ping. (2007). Spectral analysis and denoising of MIMU raw measurement. *Optics and Precision Engineering*, Vol.15, No.2, (February 2007), pp. (261-266), 10042924X(2007) 0220261206.

Mallat S, Hwang W L. (1992). Singularity detection and processing with wavelets. *IEEE Transactions on Information Theory*, Vol.38, No.2, (March 1992), pp. (617–643). 0018-9448

Napolitano M R, Windon D A, Casanova J L, Innocenti M, Silvestri G. (1998). Kalman filters and neural-network schemes for sensor validation in flight control systems. *IEEE Transactions on Control System Technology*, Vol.6, No.5, (September 1998), pp.(596–611), 1063-6536.

Qi Juntong, Han Jianda. (2007). Application of Wavelets Transform to Fault Detection in Rotorcraft uav sensor failure. *Journal of Bionic Engineering*, Vol.4, No.4, (December 2007) pp.265-270, 10.1016/S1672-6529(07)60040-7.

Qi Juntong, Song Dalei, Dai lei, Han Jianda. (2010). The ServoHeli-20 rotorcraft UAV project. *International Journal of Intelligent Systems Technologies and Applications*, Vol.8, No.1-4, (2010) pp.(57-69), 10.1504/IJISTA.2010.030190.

Qi Juntong, Zhao Xinggang, Jiang Zhe, Han Jianda. (2006). Design and implement of a rotorcraft UAV testbed. *IEEE International Conference on Robotics and Biomimetics*, 1-4244-0571-8/06, Kunming, China, December 2006.

Qu Guofu, Zhao Fan, Liu Guizhong, Liu Hongzhao. (2009). Adaptive MEMS Gyroscope Denoising Method Based on the à Trous WaveletTransform. *The Ninth International Conference on Electronic Measurement & Instruments*, 978-1-4244-3864-8/09, Beijing, China, February 2009.

Rudin W. (1976). *Principles of Mathematical Analysis* (Third), McGraw-Hill, 007054235X, New York.

Shi Yu, Zhang Xianda. (2000). Kalman-filtering-based angular velocity estimation using infrared attitude information of spacecraft. *Optics and Precision Engineering*, Vol.39, No.2, (February 2000), pp. (551-557), 10.1117/1.602394.

Song Lijun, Qin Yongyuan,Yang Pengxiang. (2009). Application of Wavelet Threshold Denosing on MEMS Gyro. *Journal of test and measurement technology*, Vol.23, No.1, (April, 2008), pp.(33-36), 1671-7449(2009)01-0033-04.

Su Li, Zhou Xue-mei. (2009). Application of improved wavelet thresholding method for de-nosing gyro signal. *Journal of Chinese Inertial Technology*, Vol.17, No.2, (April 2009), pp.(231-235), 1005-6734(2009)02-0231-05.

Wu Chong, Song Dalei, Dai Lei, Qi Juntong, Han Jianda,Wang Yuechao. (2010). Design and Implementation of a Compact RUAV Navigation System. *IEEE International Conference on Robotics and Biomimetics*, 978-1-4244-9319-7, Tianjing, China, July 2010.

Zhang Jianqiu, Ma Jun, Yan Yong. (2000). Assessing blockage of the sensing line in a differential-pressure flow sensor by using the wavelet transform of its output. *Measurement Science and Technology*, Vol.11, No.3, (March 2000), pp.(178–184),10.1088/0957-0233/11/3/302.

Zhang Jianqiu, Yan Yong. (2001). A wavelet-based approach to abrupt fault detection and diagnosis of sensors. *IEEE Transaction on Instrumentation and Measurement*, Vol. 50, No.5, (October 2001), pp.(1389–1396), 0018-9456.

Permissions

The contributors of this book come from diverse backgrounds, making this book a truly international effort. This book will bring forth new frontiers with its revolutionizing research information and detailed analysis of the nascent developments around the world.

We would like to thank Dumitru Baleanu, for lending his expertise to make the book truly unique. He has played a crucial role in the development of this book. Without his invaluable contribution this book wouldn't have been possible. He has made vital efforts to compile up to date information on the varied aspects of this subject to make this book a valuable addition to the collection of many professionals and students.

This book was conceptualized with the vision of imparting up-to-date information and advanced data in this field. To ensure the same, a matchless editorial board was set up. Every individual on the board went through rigorous rounds of assessment to prove their worth. After which they invested a large part of their time researching and compiling the most relevant data for our readers. Conferences and sessions were held from time to time between the editorial board and the contributing authors to present the data in the most comprehensible form. The editorial team has worked tirelessly to provide valuable and valid information to help people across the globe.

Every chapter published in this book has been scrutinized by our experts. Their significance has been extensively debated. The topics covered herein carry significant findings which will fuel the growth of the discipline. They may even be implemented as practical applications or may be referred to as a beginning point for another development. Chapters in this book were first published by InTech; hereby published with permission under the Creative Commons Attribution License or equivalent.

The editorial board has been involved in producing this book since its inception. They have spent rigorous hours researching and exploring the diverse topics which have resulted in the successful publishing of this book. They have passed on their knowledge of decades through this book. To expedite this challenging task, the publisher supported the team at every step. A small team of assistant editors was also appointed to further simplify the editing procedure and attain best results for the readers.

Our editorial team has been hand-picked from every corner of the world. Their multi-ethnicity adds dynamic inputs to the discussions which result in innovative outcomes. These outcomes are then further discussed with the researchers and contributors who give their valuable feedback and opinion regarding the same. The feedback is then collaborated with the researches and they are edited in a comprehensive manner to aid the understanding of the subject.

Apart from the editorial board, the designing team has also invested a significant amount of their time in understanding the subject and creating the most relevant covers. They scrutinized every image to scout for the most suitable representation of the subject and create an appropriate cover for the book.

The publishing team has been involved in this book since its early stages. They were actively engaged in every process, be it collecting the data, connecting with the contributors or procuring relevant information. The team has been an ardent support to the editorial, designing and production team. Their endless efforts to recruit the best for this project, has resulted in the accomplishment of this book. They are a veteran in the field of academics and their pool of knowledge is as vast as their experience in printing. Their expertise and guidance has proved useful at every step. Their uncompromising quality standards have made this book an exceptional effort. Their encouragement from time to time has been an inspiration for everyone.

The publisher and the editorial board hope that this book will prove to be a valuable piece of knowledge for researchers, students, practitioners and scholars across the globe.

List of Contributors

Bouden Toufik
Automatic Department, Laboratory of Non Destructive Testing, Jijel University, Algeria

Nibouche Mokhtar
Bristol Robotic Laboratory, Department of Electrical and Computer Engineering, University of the West of England, UK

Pooneh Bagheri Zadeh
Staffordshire University, UK

Akbar Sheikh Akbari and Tom Buggy
Glasgow Caledonian University, UK

Matej Kseneman
Margento R&D d.o.o., Slovenia

Dušan Gleich
University of Maribor, Faculty of Electrical Engineering and Computer Science, Slovenia

Ehsan N. Arya, Z. Jane Wang and Rabab K. Ward
The University of British Columbia, Vancouver, Canada

Burhan Ergen
Fırat University, Turkey

Munesh Chandra
DIT School of Engineering, Greater Noida, India

Daniel Acevedo and Ana Ruedin
Departamento de Computación, Facultad de Ciencias Exactas y Naturales, Universidad de Buenos Aires, Argentina

Shahid M. Satti, Leon Denis, Ruxandra Florea, Jan Cornelis, Peter Schelkens and Adrian Munteanu
Department of Electronics and Informatics (ETRO), Vrije Universiteit Brussel-IBBT, Brussels, Belgium

Ching-Yu Yang
Department of Computer Science and Information Engineering, National Penghu University of Science and Technology, Taiwan

Guangyu Wang, Qianbin Chen and Zufan Zhang
Chongqing Key Lab of Mobile Communications, Chongqing University of Posts and Tele-communications (CQUPT), China

Tomoki Ikoma, Koichi Masuda and Hisaaki Maeda
Department of Oceanic Architecture and Engineering, College of Science and Technology (CST), Nihon University, Japan

Juntong Qi and Jianda Han
State Key Laboratory of Robotics, Shenyang Institute of Automation, Chinese Academy of Sciences, People's Republic of China

Lei Dai and Chong Wu
State Key Laboratory of Robotics, Shenyang Institute of Automation, Chinese Academy of Sciences, People's Republic of China
Graduate School of Chinese Academy of Sciences, People's Republic of China